PC Interfacing for Data Acquisition and Process Control

2nd Edition

Sanjay Gupta

&

Jai P. Gupta

PC INTERFACING FOR DATA ACQUISITION AND PROCESS CONTROL   2nd edition

Printed in the United States of America

In preparing this work the author and publishers have not investigated or considered patents which may apply to the subject matter hereof. It is the responsibility of the readers and users of the subject matter to protect themselves against liability for infrengement of patents. The information and recommendations contained herein are not intended for specific applications but are of a general educational nature. Accordingly, the authors and publishers assume no responsibility and disclaim all liability of any kind, however arising, as a result of using the subject matter of this work.

All trademarks referenced in this book are trademarks or registered trademarks (as applicable) of their respective owners.

The equipment referenced in this work has been selected by the author as examples of the technology, no endorsement of any product is intended by the publisher. In all instances, the manufacturer's procedures should prevail regarding the use of specific equipment. No representation, expressed or implied, is made with regard to the availability of any equipment, process, formula or other procedures contained herein.

While the statements and comments in each chapter are based on each individual author's experience and views, none are to be construed as recommendations of one product over another.

The computer codes and laboratory exercises given have been tested by the respective authors. However, no resonsibility will be accepted by author(s)/editors/publisher/ distributors/sellers of this book for any loss of life, damage to any living being and/or material, or loss of time while carrying out such exercises or using the detailed software. Total care must be exercised by the user to ensure proper use.

INSTRUMENT SOCIETY OF AMERICA
67 Alexander Drive
P.O. Box 12277
Research Triangle Park, NC 27709

Library of Congress Cataloging in Publication Data

PC interfacing for data acquisition and process control /
[edited by] Sanjay Gupta & Jai P. Gupta. — 2nd ed.
p. cm.
Includes bibilographical references and index.
ISBN 1-55617-513-2
1. Process control—Automation. 2. Automatic data collection systems.
3. Computer interfaces. 4. Microcomputers. I. Gupta, Sanjay.
II. Gupta, J. P. (Jai P.)
TS156.8.P38 1994
670.42'75—dc20

94-10045
CIP

To Ira and Prabha

Table of Contents

Preface to the Second Edition

Since the publication of the first edition of *PC Interfacing for Data Acquisition and Process Control* five years ago, phenomenal progress in the world of computers has been made. While the principles of interfacing have remained more or less the same, the practice has become far more sophisticated and widespread. This edition reflects the progress over the past five years.

Chapter 1, "The Personal Computer, and MSDOS," has been greatly expanded and totally rewritten to reflect the current state of the hardware and software. Chapter 3 has been replaced with a new chapter on "Software: Tools and Techniques." This chapter not only deals with software tools in detail but also introduces the area of object-oriented programming (OOP) methods at some length. The techniques of OOP have taken the world by storm, but many potential users are put off by the "hype" surrounding what is a very logical and simple extension of normal structured programming. Understanding of OOP techniques can pay very rich dividends in the interfacing world. Chapter 4, "Language Options," has been extensively revised to reflect the current trends in programming style and choice.

Chapter 6, "Buses and Communications," has been almost totally rewritten. While five years ago networking was a skill required and practiced by few, today it is a way of life. Staring from small local peer-to-peer LANs, it extends all the way to global systems (such as Internet) that link millions of users around the globe. A basic understanding of the principles of networking is essential in order to take full advantage of the facilities of the "Global Village" of today. Chapter 8, "Interfacing: Digital and Analog," has been updated to reflect new developments as well as the experience of the author. Chapter 9, "Interfacing the Personal Computer," has been expanded to include discussion of the hardware and software of the various AT-class machines, in addition to the PC class systems that are the basis of the entire family. Chapter 10, "Automated Laboratory," reflects the current thinking on laboratory scale automation. Once again, the thinking on the subject has developed and matured in the past five years. The scope of Chapter 11, "Real Life Applications and Caution," has been expanded to consider various safety considerations in modern day system design and implementation. Various disasters, such as those at Bhopal, Chernobyl, Three Mile Island, and Exxon Valdez and with the Space Shuttle, have once again shown that the cost of compromising on safety can be very high.

We hope our readers will like this new edition of the book and look forward to feedback from them in this regard.

Preface to the First Edition

The introduction of computers in various aspects of data collection, control, and analysis over the past few years has revolutionized modern day research and development and manufacturing. The scientist working in a laboratory often feels lost and helpless in the world of computerization and automation. The widespread belief that laboratory interfacing requires expert knowledge about computer hardware and assembly language programming deters him from venturing into this new world. Today this is no longer true as there are excellent ready-made hardware and software packages that allow him to make maximum use of the computer with a minimum of effort. The specialized knowledge required to venture into the world of laboratory automation is minimal. This book is targeted at the user who wishes to use computers in his laboratory or industrial manufacturing activity. It contains the necessary information to enable him to do so, not only with bus-based instruments but also with older instruments, and to set up automated systems under computer control.

Over the past few years the IBM-PC™ has emerged as a leader in the world of small computers. An entire industry has developed around the PC for supplying various add-on cards and software, permitting its use as a very powerful system for data acquisition and control. Unfortunately, most of the books in the area are either aimed at the computer professional or the absolute novice. The intelligent experimenter finds himself at a loss for suitable books covering his needs. The authors faced this problem during their attempts to use the PC in the laboratory. This book is based on the experience of using PCs in the laboratory environment for several years. The material has also been used in an intensive course on Laboratory Automation for engineering college teachers, R & D scientists, and industrial users.

Historically, the first computer to make a significant impact on data acquisition and control was the legendary PDP-8™, which in its several variants was used extensively in major laboratories all over the world. However, cost and complexity precluded its use for all but the most sophisticated instruments. The first microprocessor-based system (personal computer) to have a significant impact in the laboratory was the Apple™, which was based on the 6502 processor. For the first time the user was provided with a system that was cheap, reliable, had ample computing and graphics capabilities, and was easy to use. Its open bus structure with a high degree of standardization led to the growth of an industry to exploit the Apple for the laboratory through interface cards, etc. Even today a vast array of commercially available research equipment has been designed to interface with the venerable Apple. The Intel™ 8080 family (8080, 8085 and Zilog Z80™) based computers never found widespread use in the laboratory due to the lower level of standardization in the hardware and firmware and the almost total absence of graphics. On the industrial front, the STD bus is by far the most popular. The PC family in a ruggedized form is becoming increasingly popular. It is expected that the PC/AT™

and the VME™ systems shall be the standards of the future. While the former will find use in smaller applications, the latter is expected to dominate the segment currently held by small- and medium-sized mainframes.

With the arrival of the IBM-PC, a far more powerful machine with good graphics capabilities and excellent though conservative hardware design became available. The reputation of IBM, coupled with the excellence of its design, led to its very rapid acceptance. The development of MSDOS™ version 2 also made it very easy to add and integrate additional hardware devices, which could be run with minimal software effort. Today the commercially available hardware makes it unnecessary for the user to know assembly language programming, and he can happily use the PCs in BASIC, PASCAL, FORTRAN, etc., using the available handlers. The interrupt structure of the PC is very well designed and permits the user to make full use of the power of the machine for interfacing.

This book aims to provide the user with a concise collection of information that will enable him to use the PC in his everyday work in an intelligent and successful manner. The reader is expected to be familiar with at least one high-level programming language like BASIC or PASCAL. No attempt is made to teach the reader a programming language or MSDOS. Excellent texts are available for these.

Chapter 1 introduces the personal computer and its operating system MSDOS™ along with some details of the hardware and software. Simple information regarding the choice of systems and options for laboratory interfacing applications is also provided.

Chapter 2 is devoted to software design and data structures. For successful data acquisition and control the design of the data structure and the software is critical. By its very nature software intended for use in an R & D environment has to undergo frequent changes. Quite often these modifications are evolutionary. The same basic software may be used for a long time by different users, or it may have to be modified slightly for a different application. Modular and structured design is therefore absolutely essential. Often, many modules share the same utilities; therefore, good data structure becomes mandatory.

Software tools are discussed in Chapter 3. There is a vast amount of software available for the PC. To the user a computer is only as good as the software available for it. When the PC was introduced, it was by no means the most powerful personal computer on the market. What led to its immense success was the software available for it. This chapter discusses the various types of software available for the PC and how it can be used for laboratory purposes.

Chapter 4 deals with the language options. The choice available to a user is indeed vast. On the one hand, there is interpreted BASIC, which is included with the operating system. On the other hand, there are packages like ASYST, which offer very powerful features for data acquisition and analysis at a matching price. This chapter deals with the range of the languages and other software available for data acquisition and analysis and the advantages and disadvantages thereof.

Chapter 5 introduces the essentials of sensors and associated systems. Sensors and transducers are the eyes and ears of any real-life interface. The parameter under study is detected by a sensor, and the control system, in turn, actuates the control element. For good performance the selection of a proper sensor is of paramount importance. In addition to describing some common types of sensors, various considerations (e.g., grounding, avoiding noise loops, etc.) required for successfully interfacing the sensors to the controller are also discussed.

Buses and communications are discussed in Chapter 6. Networking and machine-to-machine communication are becoming increasingly important in the field of computer applications. Almost every computer you may care to buy has at least a serial (RS-232C) and a parallel port available. A large number of stand-alone analytical instruments provide an RS-232C interface for adding a control computer, which significantly enhances the capabilities of the system. The various ways of configuring the RS-232C are discussed. This chapter also discusses the common LANs Ethernet™ and MAP.

Chapter 7 is devoted to GPIB™ or IEEE 488 interfacing. GPIB for many people is synonymous with laboratory automation. A vast array of testing and measuring instruments are available that can be interconnected through GPIB. This chapter also explains the operation of the GPIB interface and how it is commonly implemented on the PC.

Chapter 8 concerns digital and analog interfacing. The real world is analog, and it is always not practical to use expensive conditioners/controllers to interface the signal digitally to the computer through GPIB or RS-232C. The most powerful and yet often neglected area of laboratory automation is that of analog and digital interfacing. Digital interfacing is extremely useful for sensing and activating elements like valves, motors, solenoids, etc. This chapter discusses the various ways in which, using a typical ADC-DAC-DIO card, transducers can be interfaced directly with the computer and control elements activated.

Chapter 9 discusses interfacing the IBM PC family: the requirements and restrictions of hardware, software, input-output addressing, interrupts, etc. Some knowledge of these is essential for successfully integrating commercially available interface cards with the PCs and also for utilizing their full capabilities.

Chapter 10 is devoted to the automated laboratory. Many practical situations are discussed. Some techniques for using apparatus not belonging to the computer age are also discussed. When an experimenter starts out to acquire a system for the laboratory, there is a basic choice to be made between an automatic “push button” instrument and a system assembled from various modules. The options and their merits are discussed. These would also apply to the development of custom-made systems for specific applications. The requirements of the human interface for software intended for use in the Automated Laboratory are also discussed.

Chapter 11 discusses some real-life applications where the techniques discussed in the earlier chapters are applied. It illustrates how a PC-based system can be used in the laboratory and in industry. Examples of control of a batch distillation column, bioreactor, and furnace are presented. Application of small computers in a networked environment for the monitoring and control of a long pipeline are also discussed. This chapter also introduces some concepts of safe or error-tolerant operation.

Chapter 12 gives some laboratory exercises designed to initiate the user into the fascinating area of computer interfacing. The Turbo PASCAL™ or BASICA™ programs are also given with each exercise to enable the user to study and understand good programs in interfacing. The exact nature of the instruments vary widely from place to place, but the exercises given here will serve as a very useful training aid for interfacing. The exercises and variations on them can be used to design a course on laboratory automation. An illustrative example of writing a handler in a high-level language for an interface card is also given with a detailed program listing. This handler performs most of the functions performed by the card manufacturer’s own handler. All the exercises have been used with advantage in various courses.

The *Glossary* contains an extensive listing with brief explanations of the various terms used in the book and should help familiarize the reader with much of the jargon of this fascinating area.

Throughout the book the emphasis is on applications rather than on theory, on which several books are available. Every good experimenter develops a feel for the job at hand, and this book should help him to successfully use his experience in automating his laboratory. The editors have set up many laboratory systems of differing complexity without having to resort to any machine language programming; this should serve to convince the experimentalist that he can go a long way by making full use of the existing tools.

It is hoped that this book will help dispel the false notion that interfacing computers is a black art and encourage many readers to benefit from the advantages of fully exploiting the potential of small computers in the laboratory or plant.

1
The Personal Computer and MSDOS

Sanjay Gupta

1.1 Introduction

The world of personal computing was dominated by Apple® IIs and CPM-based machines until 1981 when the giant of the computer industry, International Business Machines (IBM)® announced its Personal Computer (PC) to the small systems market. This machine was a major deviation from the norms of the computer industry in that most of the details of the system were disclosed. Even the source code of the basic input-output system (BIOS), though copyrighted, was published. The next landmark came a year later when Compaq® announced the first clone of the PC. The availability of low-cost clones led to an avalanche of PCs of varying price, performance, and quality. Initially, the degree of compatibility of the clones varied, but over a period of time compatibility has ceased to be an issue. IBM followed the PC with the AT in the last quarter of 1984, using the 80286 processor and a bus augmented to support 16-bit operations. This bus later came to be known as the Industry Standard Architecture (ISA bus). In 1986 the first systems using the 80386 CPU came not from IBM but from clone makers.

In 1987 IBM announced the PS/2 family to replace the PC. While the bottom end of the PS/2 range was very similar to the PC and used the same bus, the top end used a different bus architecture known as micro channel architecture (MCA). These machines did not take the market by storm as IBM had hoped, and ultimately IBM was forced to reintroduce the ISA bus in the ValuePoint range in 1992.

With rapid standardization of components and subsystems, the quality of clones has improved tremendously. Today, almost all clones offer reasonable quality and performance. As the market has become increasingly competitive, major manufacturers have been forced to offer low-cost machines in addition to their regular systems. The major differences between the more expensive and cheaper systems are more in the form of brand image, after sales service, expansion capabilities, etc., than in basic performance and reliability. Only at the very top end do the premium machines offer major performance advantages over what are often referred to as "no name" clones.

Today, there is vast variation in machines available to the user, ranging from low-cost PC-XTs (now virtually obsolete) to Pentium™-based machines with performances well into the minicomputer range. This chapter introduces the basic hardware and software of the PC family along with the details regarding the requirements for a system for use in data logging and process control. The various specialized interface cards are discussed in later chapters.

1.2 The Hardware

The hardware is best examined by looking at the different aspects; namely, CPU, bus, graphics, peripherals, and the full system. With the very widespread use of networking, one may also say that networking hardware should also be included. This will, however, be considered in a later chapter.

1.2.1 The CPU

The PC family is based on the Intel® 80xxx family of CPUs. The original PC used the 8/16-bit 8088 processor from Intel. Intel called it an enhanced 8-bit processor since it had an 8-bit bus supporting a 16-bit CPU. The 8088 has evolved into a whole family of processors. The major landmarks in this evolution have been the 80286, 80386, 80486, and the Pentium. The 80286 was a 16-bit processor used in the original AT. The 80386 is a 32-bit processor. Later a version of 80386, using a 16-bit external bus (to enable it to use peripheral chips of the 286 machines) and known as the 80386SX, was announced. The 32-bit bus version of 80386 was subsequently renamed the 80386DX. All these processors have associated numerical coprocessors (8087, 80287, 80387SX, 80387DX) for improved floating point performance. The 387 class coprocessors have built-in transcendental functions. The 80287 has had this only since late in 1990. While the 8088-8087 combination typically gave a 10 times improvement in performance, the better designed CPUs of the 80286 and 386 classes gain only a 3× improvement from the coprocessor.

Intel then combined the 80386 and 80387, along with cache management and an 8-kB on-board cache, into a single chip named the 80486DX. A stripped-down version (without the coprocessor) known as the 80486SX was also released. An 80487SX numerical coprocessor soon followed. The numerical coprocessor has evolved into the overdrive family of coprocessors, which provides enhanced performance for some specialized applications (such as CAD) or permits retrofitting of the higher processors through the coprocessor socket. The Pentium 32-bit CPU followed, providing another quantum leap in performance. Intel deviated from the numerical numbering for the Pentium since, while a name can be copyrighted, a numerical designation cannot. In addition to these, there are also low-power versions of the chips (typically carrying the letter L somewhere in the suffix) for use in lap top or hand- held systems.

Also, as clock speeds kept on increasing, the design of motherboards became increasingly difficult. While the original PC worked at 4.77 MHz, clock speeds of up to 50 MHz are now common. When the 50 MHz version of the Intel 80486DX was released, this limitation showed up. 50 MHz is well into the VHF range of TV transmission, and the parallel tracks on PCBs (which look so beautiful) start radiating large amounts of energy. This made upgrading of motherboards beyond 33 MHz an expensive process. Taking advantage of the 8-kB on-board cache, Intel then introduced its clock doubled (DX2) range of 486 class CPUs. For example, the 80486DX2-66 operates internally at 66 MHz but operates at 33 MHz on the bus. This way it is able to achieve about 60–80% of the performance of a pure 66 MHz design without necessitating an expensive redesign of the motherboard. In fact, but for the higher heat dissipation, the DX2 is a drop-in replacement for the corresponding DX chip. DX4 processors (actually clock trebled CPU's) have also been announced.

In the beginning, other manufacturers restricted themselves to licensed production of Intel CPUs. Innovation was initially restricted to the coprocessor market, where clone makers CYRIX® and IIT® introduced coprocessors with improved performance. Weitek® introduced array processors (3167, 4167), which have no Intel equivalent. CYRIX 80387EMC combines the 80387 with an array processor in a single chip. Over a period of time other manufacturers diversified in the area of CPUs as well. Many started offering improved performance (e.g., NEC® V20 and V30 clones of the 8088 and 8086), faster clocks (Harris® 80286, AMD® 80386 and 486). Recently, a lucrative market for pin-compatible CPUs has evolved. The prime example of this are the CYRIX Cx486SLC and Cx486DLC. Both support the 486SX instruction set with 1 kB of on-board cache. The former is pin-compatible with the 386SX and the latter with the 386DX. IBM, under cross-licensing from Intel, released the 386SLC, which is essentially a 486SX in a 386SX package. Low-voltage (3.3 V) variants for battery-powered systems have also made their appearance.

The CPUs can essentially be divided into the following classes:

1. *8088/8086:* Used on the original PC. Now obsolete. NEC V20 and V30 are popular improved performance variants.

2. *80286:* Used in the original AT. Supported a 16-bit bus with 24-bit address. Protected mode allowed it to address memory beyond 1 MB. The protected mode was, however, flawed and only limited use of it could be made. Faster 80286 CPUs were released by Harris and others. The 80287 initially had essentially an 8087 instruction set but was later upgraded to the 80387 set, which includes transcendental functions.

3. *80386:* A 32-bit CPU with a 32-bit bus. The 80386SX supports a 16-bit bus and has been rightly called the *processor that the 80286 should have been but never was*. Full protected mode operation is supported. The SX can support only up to 16 MB (24 bits) of physical memory. Virtual memory up to 4 GB (2^{32} bytes) is supported by all chips. In the protected mode, 386 (and higher class) machines allow the user to write programs using memory of up to 4 GB. The use of DOS extenders and appropriate compilers, however, becomes necessary.

4. *80486:* Combines the 80386, a cache controller, 8-kB on-board cache, and the 80387 (not in 80486SX) in a single chip with additional optimization. The CYRIX 486 and IBM 386SLC fall somewhere in between the 386 and 486SX in performance.

5. *Pentium:* A 32-bit CPU with vastly improved performance. It has 64-bit data and address buses with 16-kB cache and a highly optimized pipelined design.

1.2.2 The Bus

The system bus of the PC was an 8-bit bus that used a 62-pin edge connector. This was later upgraded to a 16-bit bus on the AT by the addition of another connector. This has come to be referred to as the ISA (Industry Standard Architecture) bus. IBM then announced the PS/2 range, using the Micro Channel Architecture (MCA) with totally different connectors, with support for advanced features like multiple masters, burst mode operation, etc. This architecture was not available in the public domain to others. The industry led by the so-called Gang of Nine (Compaq®, Olivetti®, NEC®, AST®, Hewlett-Packard®, Epson®, Tandy®, Wyse®, and Zenith®) responded by announcing EISA (Extended ISA), with similar capabilities. The EISA connector is similar to the ISA but deeper supporting another set of pins at the bottom. ISA boards will work happily with the EISA machines. In the permanent quest for better performance machines, auxiliary buses that directly interface with the CPU have made their appearance. It may be kept in mind that the original PC bus is of 1981 vintage, and it is not likely to be suitable for the technology of the 1990s. Systems based on the Video Equipment Standards Association (VESA) Local Bus (VL-Bus) are becoming increasingly popular. Intel has defined the peripheral component interconnect (PCI) standard. While systems up to the 486 class machines work fine with the VL-bus, Pentium class machines need the PCI bus for optimum performance. The ISA bus can operate at rates of about 5 MB per second. MCA and EISA increase this rate to 32 MBps, while PCI and VL bus can operate up to 130 MBps. Apple has recently introduced the QuickRing™, which is capable of operating at speeds of up to 350 MBps. At the same time the PC bus which initially ran at 4.77 MHz was speeded up to a de-facto standard of 8 MHz in ISA, it has been further accelerated to 10 MHz. Thus, peripheral cards that ran on earlier machines may not necessarily run reliably on later machines.

The choice available to the user who is buying an ISA-based system is whether to go for plain ISA or also the EISA and/or VESA extensions (PCI is being used in some Pentium systems). The main limitation of the ISA design is that the system cannot address more than 16 MB of RAM; EISA does not have this limitation. EISA and VESA are beneficial only if the system takes advantage of these extensions. The number of unsuspecting buyers of EISA systems with no EISA peripherals and less than 16 MB RAM is phenomenally high. Thus, the buyer should not spend extra money on EISA (and VESA) unless there is a need for it.

1.2.3 The Graphics System

The quality of displays on PCs has improved greatly from the original CGA to the VGA, SVGA, and XGA standards. The biggest contribution of the standards from VGA onwards has been the *square pixel.* The screen has an aspect ratio of 4:3. Graphic standards prior to the VGA (Hercules, CGA, EGA) did not have their number of pixels in the same ratio, making it necessary to correct the pixel ratio in order to display objects in their proper shape. VGA, with 640×480 pixels, finally made the display and pixel ratio the same. For example, to draw a square, 100×100 pixels will appear as a square in VGA (and later) but will appear as a rectangle in the earlier standards. The various standards have been as follows:

1. *Color Graphics Adapter (CGA):* It had text in an 8×8 matrix. Text could be displayed in 40×25 or 80×25 formats with graphics in 320×200 (16 colors) or 640×200 (2 colors) resolution. It is now obsolete. Windows™ version 3.1 and later do not support the CGA.

2. *Hercules Graphics Card (HGC):* Hercules took the 14×9 matrix of the original monochrome display adapter and extended it into a 720×348 monochrome display. This standard was not directly supported by IBM but nevertheless became very widely used, and many clones appeared. It is also known as the "monochrome graphics (or display) adapter" (MDA). Combined CGA/MDA cards were also very popular.

3. *Enhanced Graphics Adapter (EGA):* This graphics card was an enhancement of the CGA with a resolution of 64×350 in 16 colors. All EGA cards support the CGA modes and most also support the HGC modes.

4. *Video Graphics Array (VGA):* This has a resolution of 640×480 with 16 colors. The palette of 16 colors can be selected from a set of 256 colors. The VGA also moved from digital (TTL) to analog signals. The VGA also supports all EGA (and earlier) modes. VGA cards have memory of 256 kB or more.

5. *Super VGA (SVGA):* These are enhanced VGA boards that support higher resolution modes of 800×600, 1024×768, and so on. These have a memory of 512 kB, 1 MB, or more. To take advantage of the higher resolution modes, graphics drivers specific to the boards are normally required. However, the 800×600 mode has been standardized by VESA, and generic drivers work very well. Video accelerator boards (using special on-board CPUs and special software drivers) offer enhanced performance. Most of these also offer the same resolutions as SVGA boards.

In addition, many other standards, such as extended CGA (offering 16 colors at 640×200), Hercules Color Card (HGC in color), IBM 8514, IBM Professional Graphics Adapter (PGA), and IBM Extended Graphics Array (XGA), have also appeared with varying popularity.

1.2.4 The Peripherals

A motherboard and graphics system alone cannot become a complete system. Other peripherals, such as disk drives (floppy as well as hard), keyboard, mouse, interface cards, etc., are also required. One can consider the graphics system as a peripheral as well. The first peripheral to be considered is the auxiliary storage. This can be in the form of floppy and hard disks, tapes, optical storage devices, CD ROMs, and the like. Let us consider these in turn.

1.2.4.1 Floppy and Hard Disks

Magnetic disks almost invariably form the main auxiliary storage of PCs. The software being used is stored on these devices. Modern software has become too large and complex to permit meaningful usage from floppy diskettes. Floppy diskettes come in two sizes: 5-1/4 in. and 3.5 inches. The latter is a newer medium that has rapidly gained popularity since 1989. 5-1/4-in. diskettes for the PC come in two capacities: 360 kB and 1.2 MB. The PC under DOS 1.1 used a capacity of 320 kB, but that became obsolete a decade ago. While the PC used 360-kB drives, AT and higher machines used 1.2-MB drives. The 1.2-MB drives can read and write 360-kB drives. However, this cannot always be relied upon. The 1.2-MB drive obviously puts more than three times the information on the same area of the disk. To achieve this, not only has the writing density been increased from 9 sectors to 17 sectors, but also the track spacing has also been reduced from 48 tracks per inch (tpi) to 100 tpi. By combining these, the disk capacity has increased from 9 sectors × 40 tracks × 2 heads × 512 bytes or 368,640 bytes to 17 sectors × 80 tracks × 2 heads × 512 bytes or 1,392,640 bytes. The increase in track density requires a narrower head on the drive. This sometimes leads to problems in reading and writing diskettes of 360 kB on 1.2-MB drives. The 3.5-in. medium was adapted on the PC at the time of the release of the PS/2. These diskettes are smaller and more robust; the plastic jacket and the cover for the head slot makes them much more rugged. Originally, the formatted capacity was 720 kB (9 sectors × 80 tracks × 2 heads × 512 bytes). The high density variant doubles the number of sectors to increase the capacity to 1.44 MB. This route has the advantage of increasing the capacity without changing the width of the heads, which is the bugbear of the 5.25-in. medium. The drives can also recognize the medium's capacity by the presence of an additional index hole in a 1.44-MB medium. However, not all PCs take advantage of this signal. The practice of punching an additional hole in a 720-kB medium to record it at 1.44 MB is not to be recommended. The high density media are not only tested at a higher capacity, but the magnetic material is also different, requiring higher current in the heads. Thus, reliable recording should not be expected. The converse practice of converting 1.44-MB diskettes to 720 kB should also be discouraged for the same reason. The drive recognizes the type of medium and adjusts the write current to the heads accordingly. Recently, 2.88-MB drives have also been introduced.

Hard disks are essential on any system. Originally, the drives were in the 5-1/4 in. form factor and used a technique known as MFM (modified frequency modulation) for storing data. This was supplemented by RLL (run length limiting), which increased disk capacity by about 50%. The ATA standard (AT Adapter) commonly called the IDE (integrated drive electronics) interface, whose advent coincided with the introduction of 3.5-in. drives, rapidly replaced the use of MFM and RLL standards. IDE drives usually have a built-in cache that permits operation at an interleave of 1:1. On the other hand, only the fastest MFM and RLL controllers were able to achieve this speed. To understand interleave, one should look at the method of reading data from the drive. The controller will read (and process) a sector. Then it transfers this to the PC memory. While this is going on, the disk is still revolving. Thus, when the controller is again free to read the next sector of data, the head would have rotated beyond the sector that was read last. Thus, if data were written on successive sectors, the controller would have to wait until the right sector again presented itself. This leads to an undue waste of time. Suppose the time taken to process and transfer corresponds to the time taken for more than one but less than two sectors to pass under the head. Then, if data were written in a way so as to sequence on every third sector, the speed would be maximum. This will correspond to an interleave of 3 (as was the case with most XTs). Most ATs and some 386-class machines could not operate with an interleave of better that 2. On IDE drives, a cache on board permits the handling of data on a track-by-track basis, allowing a far higher throughput to be achieved. And under IDE, the disk manufacturer can define heads and cylinders as he or she pleases, without the constraints imposed by the PC (and DOS), and then *translate* the track/sector address given by the machine to the actual address on the drive.

However, IDE cannot support more than two devices in most implementations. Also, as the demand for higher performance continues to increase, the SCSI (small computer systems interface) is fast becoming very popular in high end systems. This offers higher performance (especially when coupled with EISA, MCA, or local bus controllers) and can also support up to eight devices, allowing the integration of other devices such as CD ROMs, tape cartridge drives, scanners, etc. The SCSI standard has also evolved into SCSI-2 and SCSI-3. The trend towards replacing IDE by SCSI has started, and the popularity of SCSI should increase with time.

It should be noted that SCSI is implemented in PCs in two forms. In the first (and more popular) implementation, the SCSI interface mimics the IDE. The second implementation is as an add-on with a corresponding device driver. In the first implementation the system can boot from the SCSI device, and additional devices (such as WORM, CDROM, scanners, etc.) are installed through custom drivers. Although this approach is more convenient, full SCSI performance cannot be achieved due to the restrictions made by the need to mimic an IDE interface. The latter is preferred for fast EISA and local bus systems in order to get the maximum performance from the hard disks. However, in this case, a separate small hard disk is usually necessary for booting the system.

Removable hard disks are also used. These can range from what are essentially high capacity floppies (e.g., Bernoulli box) to special cradles to take normal hard disks, to special hard disks with removable media cartridges, and beyond. Also, hard disk controllers (often SCSI), which can plug into the parallel printer port, are available.

At one time a PC hard disk was 10–20 MB in capacity with an access time of >50 ms. Today, typical systems are sold with drives in the 100–250 MB range with access times in the 15 ms range. Drives of capacity in the GB range are available. Magneto-optical dives are beginning to compete with the magnetic media in performance and price.

1.2.4.2 Magnetic Tape Cartridges

In the early days of magnetic disks when capacities were in the 10–20 MB range, floppy diskettes were the popular methods of backing up data. As the capacity of hard disks increased, the use of diskettes has become very inconvenient, and magnetic tape cartridges (typically 40–120 MB) are becoming increasingly popular. They can be interfaced in many ways. There are dedicated drives that are built into the system, an approach that is not very cost-effective for smaller systems. For these systems cartridge drives which can interface through the floppy controller, or IDE controller, or the SCSI controller are available. In addition, external systems that plug into the parallel printer port are also available.

1.2.4.3 Optical Drives (WORM, CDROM, etc.)

In addition to magnetic storage, optical storage devices are also popular. The most popular are Write Once Read Many (WORM) drives and Compact Disk Read Only Memory (CDROM) drives. Both of these suffer from the limitation that information once written cannot be erased. The storage capacities are approaching and into the GB regime. WORM drives are very popular for archival purposes and can store vast amounts of data. CDROM drives are mostly preloaded with software. Only recently have WORMs in the CDROM format become available. These also, however, must be fully loaded in one go.

1.2.5 The Full System

A few years ago, a typical PC had specifications that read something like: Intel 8088 processor; 256/512 kB of RAM; two floppy drives; color graphics card with monitor; one Centronics®-compatible parallel port; and one (or two) RS232C serial ports. Today the 8088 is dead, and many popular packages will not even run on it. Most systems are sold with at least an 80386-class CPU (the 286 is on its way out), invariably have a hard disk, RAM in excess of 1 MB is standard (a lot of windows-based software will not run with anything much under 4 MB), and the graphics is almost invariably VGA or better. Color monitors are becoming increasingly popular. At the same time, this machine costs less than half of what the standard machine of five years ago cost.

The hardware inside the box of a typical PC in a mini-desktop case is shown in Fig. 1.1. A power supply (which may be rated at anywhere between 65 W and 300 W) occupies a large part of the space. Then there are the floppy and hard disk drives, a small speaker, and a few printed circuit boards (PCBs). On the back of the system are connectors for various devices.

The board mounted on the base of the housing is the motherboard (Fig. 1.2). It contains the processor, a socket for the numeric coprocessor, and some DIP switches and jumpers for configuring the system. A large part of the motherboard is dedicated to sockets (called slots) for add-on or feature cards. The number of slots in a system can vary. Some cards designed for a slimline mini-desktop cabinet have a single slot. This slot usually takes a daughterboard, which sits vertically in it. The daughter board, in turn, provides a number of slots for mounting other cards. Some of the important components of the motherboard have been identified in Fig. 1.3. The board illustrated is an ISA board. EISA boards are distinguished by deeper slots (with additional connectors on the bottom), and VESA boards have an additional connector (MCA type) in line with one or two of the ISA slots.

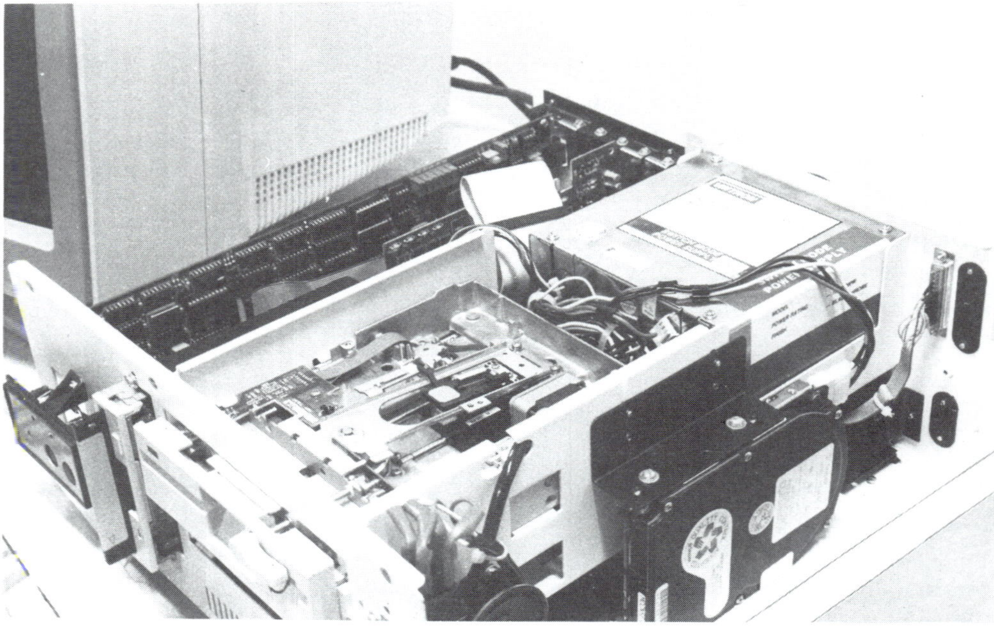

Figure 1.1—Interior of an 80486AT with Interfacing Hardware in a Mini Desktop Cabinet

(Courtesy Zenith Computers Ltd.)

The slots contain various plug-in cards. Typically, there will be a graphics card and a multifunction card. The multifunction card often includes the hard disk controller, floppy controller, serial ports, parallel port, etc., in a single card. Figure 1.4 shows such a multifunction card and also a VGA graphics card. The remaining slots are used to plug in various interfaces for data logging and control. Figure 1.5 shows an analog and digital interface card (with its screw termination panel) and an IEEE 488 interface card. Obviously, it is desirable to have a system that has a sufficient number of free slots so that the user can add cards as and when required. Other popular add-ons are networking cards, scanner cards, fax/modem cards, and so on.

1.3 The Software

A computer is of no use without software. The operating system used with the personal computer is the MicroSoft® Disk Operating System (MSDOS). IBM machines use a variant known as PCDOS. There have been six generations of MSDOS so far. Digital Research® has introduced a *work alike* called the DRDOS (now Novell DOS). In addition, Windows™ is becoming increasingly popular as a graphical user

Figure 1.2—ISA Motherboard of a 486AT Compatible

(Courtesy Zenith Computers Ltd.)

interface (GUI). The most popular (and stable) versions of MSDOS have been 2.11, 3.3, 5.0, and 6.0. The DOS diskette(s) come with the operating system and various utilities. The much maligned BASIC Interpreter™ (GWBASIC) has given way to QuickBASIC™ in DOS 5.0. IBM has introduced Operating System 2™ (OS/2) as a multitasking operating system for the PC. Modern 386- and 486-based machines are powerful enough to support Unix, which is becoming increasingly available for PCs. However, most interfacing software is written under DOS (and, increasingly, Windows); this book confines itself to the DOS environment.

The choice of software on the PC is indeed vast; ready-made packages are available for most applications one can think of. Some software may be supplied with the machine, but the quantity and quality are highly variable. For interfacing, the user will normally require at least one compiler and some other programs. A typical system used in R&D should have, in addition to the BASIC interpreter, compilers for C (MicroSoft, Borland®, Lattice®, and Aztec® are some of the more popular), FORTRAN (MicroSoft, Lahey®, and Salford® are very popular), and possibly PASCAL (almost invariably Borland) or Modula®. A word processor (MicroSoft WORD™ or WordPerfect™ are the most popular) and a spreadsheet (Lotus 123™, Quattro™, or similar) are also desirable. An assembler (such as MicroSoft MASM™ or Borland

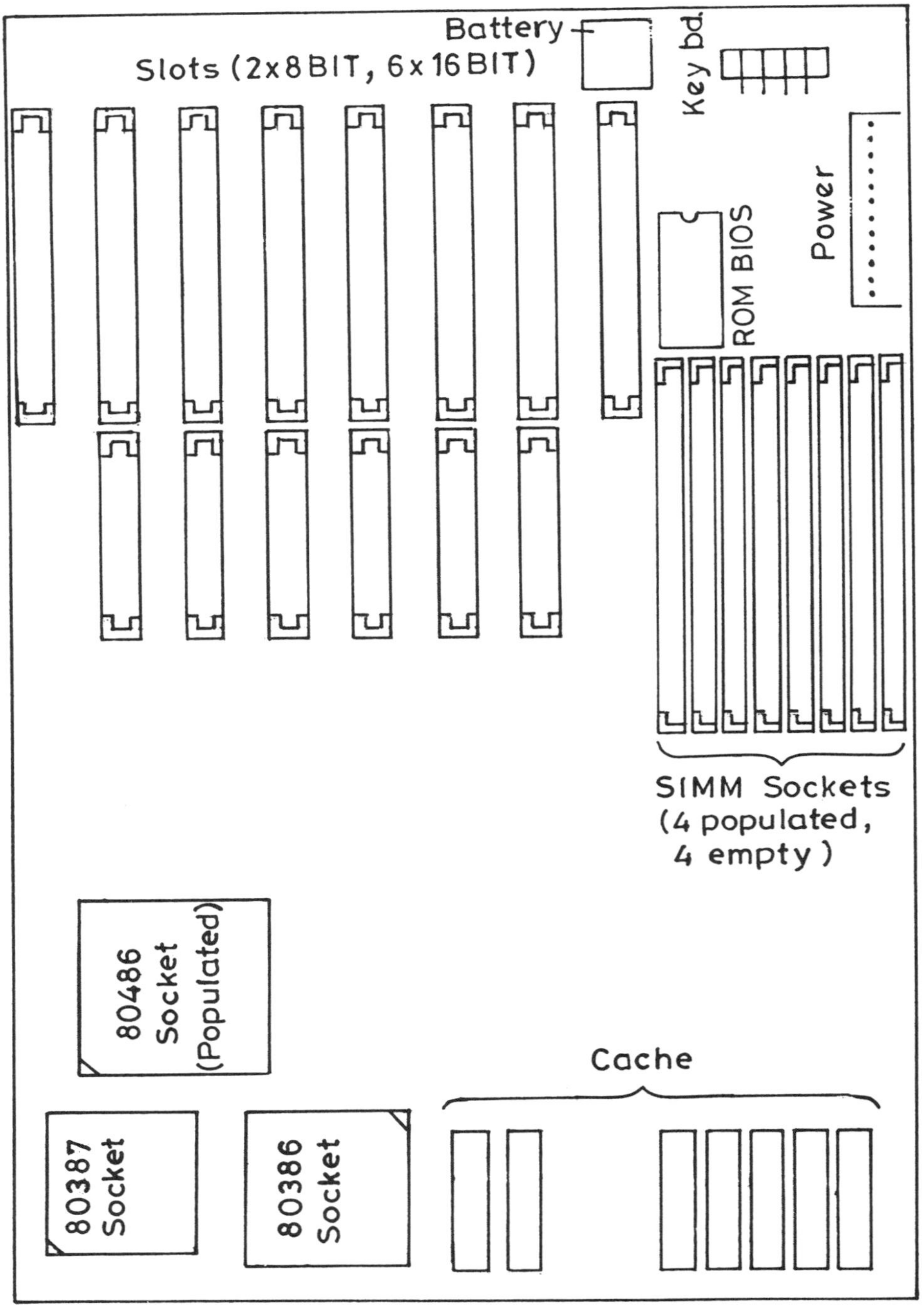

Figure 1.3—Component Overlay of ISA Motherboard

Figure 1.4—Super VGA and IDE-IO Cards for ISA Machines

(Courtesy Zenith Computers Ltd.)

TASM™) is useful in installations in which assembly-level programming is also needed. In addition, there will be software handlers for the interface cards used and sometimes specialized packages like ASYST™, Mathematica™, MathCad™, DaDisp™, and so on.

1.4 Selecting a System

The choice of a system is dictated by the often competing requirements of speed, power, and cost. Given no monetary limitations, every machine will be equipped with a superfast Pentium, with 16+ MB RAM, a few GB of storage, an ultra-high resolution graphics engine with a giant monitor, CDROM, etc., etc. However, reality

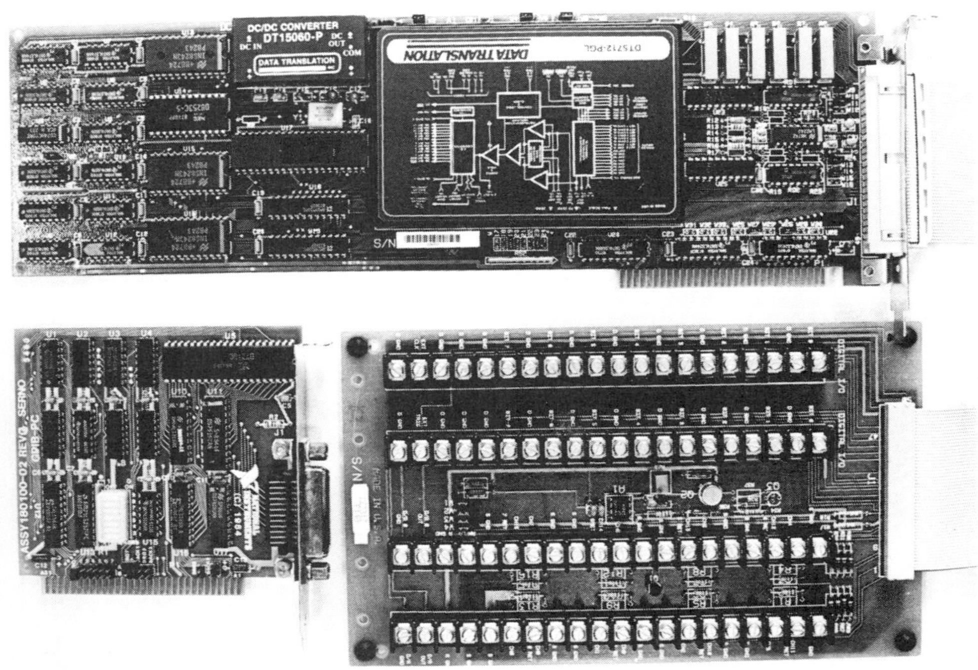

Figure 1.5—Analog & Digital Interface, and GPIB Cards

(Courtesy Data Translation Inc., National Instruments)

dictates a more realistic choice. Compatibility is no longer an issue. There are standard compatibility validation suites, and it is virtually impossible to find a system that has not been put through these. Thus, the selection essentially boils down to cost. It may be borne in mind that a far more powerful system is desirable for software development than that required for actual use. With current pricing, the minimum system that can be recommended is a 80386SX-based system with a hard disk, one floppy drive, one parallel and two serial ports, VGA graphics, and a mouse.

For interfacing work, some additional factors must be considered. The first is the power supply capacity. Any additional cards will draw power from the motherboard; thus, adequate power is necessary. The availability of expansion slots is another obvious requirement. Also, the cabinet should permit the installation of full size cards. Some ultra-compact systems may be lacking in one or more of these points. In addition, under hot conditions some systems may suffer from inadequate cooling, especially when power-hungry add-on cards are installed. Many users find mini-tower-type systems ideal for interfacing applications. The author, however, prefers desktop systems with the mini-AT form factor because they occupy much less table space.

A good graphics printer is essential. This can range from anything from an inexpensive 80-column 9-pin printer to a top-of-the-line laser. The workhorse is almost invariably a good 80-column dot matrix (24 pin) or ink jet printer. The use of plotters is decreasing because dot matrix, ink jet, and laser printers offer comparable or superior performance at a much lower cost and are also far more reliable.

Other necessities for R&D use are a good, filtered power distribution board and a sturdy trolley on wheels for the system. Quite often the computer needs to be moved close to the experimental system, and a mobile trolley is indispensable. The trolley should have adequate space for the computer system, printer, and a few instruments, if necessary. The power distribution board used for the computer should not be used for other instruments, especially if these consume large amounts of power.

1.5 Industrial PC Compatibles

The standard PC may not always be suitable for the harsh industrial environment. The electrical noise, vibration, and temperature conditions of the shop floor are very different from those of the home or the office. Various ruggedized industrial PCs, available for such applications, are of two kinds. The first kind are reworked PCs fabricated to cope with the harsher environment. These have better cooling and shielding and use assembly techniques that make them more rugged. Auxiliary storage in the form of magnetic bubble devices or static RAM banks (sometimes called silicon disks) may also be included. The second kind of system is 80x86-based and housed in standard industrial cabinets. Many buses are available for such systems, e.g., Eurobus, STD bus, MultiBus, and so on.

While systems of the first kind can be treated as PCs for most practical purposes, the second variety may range from generic MSDOS machines to close (functional) compatibles. For the first kind, standard interface cards, etc., can be used provided the cards can stand up to the environment. The level of standardization in the second kind is poor, and generally the entire system, including the add-ons, has to be procured from the same vendor. This loses much of the advantage of large-scale standardization. Even if industrial PCs have to be used in the end, it is preferable to develop the system using standard PCs and then transport to industrial PCs. If the programming is done properly, keeping the points outlined in Chapter 2 in mind, the software changes are not likely to be very extensive.

Suggested Reading

Jamsa, Kris. 1988. *DOS: Power Users Guide.* Berkeley, CA: Osborne McGraw Hill.

Murtha, S. M., and Petrie, G. R. 1983. *The PCDOS Companion.* Indianapolis: Howard W. Sams & Co. Inc.

Norton, Peter. 1986. *Inside the IBM PC.* New York: Brady/Prentice Hall.

Williams, Dave. 1990. *The Programmer's Technical Reference: MS-DOS, IBM PC & Compatibles.* New Delhi, India/London, UK: Galgotia/Sigma Press.

2
Software Design

Rajeev Sangal

2.1 Introduction

Proper software design consists of using a good data structure and developing the programs in a disciplined and structured fashion. The use of structured languages forces the developer to follow many of the good programming practices. One should be aware of the principles of good programming practice. These are outlined in this chapter. Many aspects of good programming practice come naturally if object-oriented techniques are adopted. These are discussed in Chapter 3.

2.2 Data Structure Design

The first task in software design is to identify the data that is relevant to the problem at hand. The data represents an abstraction of reality (assuming the problem pertains to reality) in the sense that certain properties and characteristics of the real objects are ignored because they are unimportant or irrelevant to the problem at hand. This is best explained by an example.

Let us say that it is desired to design software that (1) acquires data from an experiment relating to temperature and pressure of a fluid and (2) controls the flow rate by closing or opening suitable valves. Relevant parameters are the temperature and pressure. The color of the container of fluid or the material of the pipes are not important and hence can be ignored.

Having identified the data, the next task is to identify how it must be represented. The data is usually interrelated, and suitable representations must be selected for these interrelationships as well. Suppose, for example, one needs to periodically measure the temperature from three thermocouples and the pressure from two different transducers in an experiment. Moreover, the frequency of measurement of the pressures is three times that of the temperatures. The initial flow rate is given, but it must be adjusted appropriately by controlling two valves. A record must be kept of the adjustments that are done only when the value of a function applied to the temperatures and pressures crosses a certain range. In this example, temperature and pressure data and their time interrelationships are already identified. The question is: How should the data be represented?

The software must store not only the values of the measurements and the adjustments but also the relative time instants when they are made. On the one hand, the frequencies of measurements of temperatures and pressures differ by a predetermined fixed factor, and, on the other hand, the frequency of valve adjustment is also different (but not by a fixed factor). Appropriate ways must be found to keep the correspondences between the different measurements and adjustments. The following are three of several possible data structures to achieve this.

1. Maintain seven one-dimensional arrays (Fig. 2.1): three for temperatures (T1, T2, T3), two for pressure (P1, P2), and two for storing adjustments made (A, B). The *i*th index gives the respective values of the parameters at the *i*th unit or instant of time. The number of values recorded in the arrays are stored in a variable (N). The pressure measurements are made three times as often as those for temperature. Hence, there must be a way to store nil values in the arrays for temperatures when the temperature sensor is not read. One way of doing this is by storing a negative temperature, which is impossible in the absolute scale. A similar method can be used for valve adjustments. (Alternatively, the value of temperature from the previous measurement can be repeated).

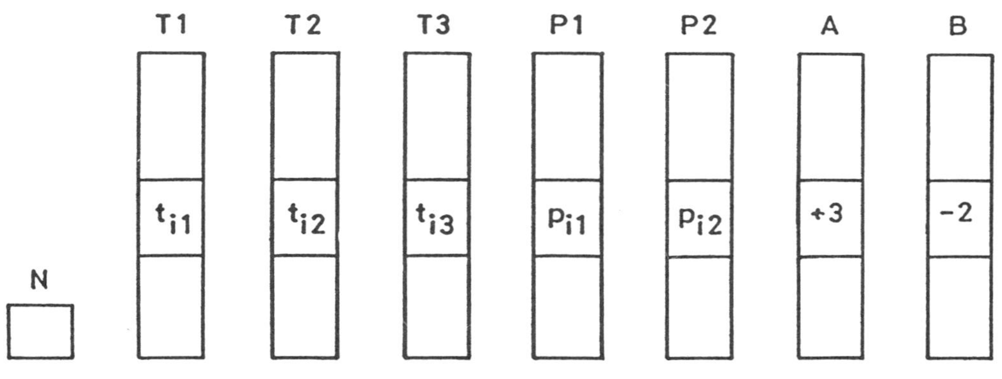

Figure 2.1—Data Structure with Seven Data Arrays

2. Maintain five one-dimensional arrays for temperatures and pressures (Fig. 2.2), where the number of values in the arrays for the temperatures is one-third that of the arrays for the pressures. Three additional arrays store the time instants and the values when valve adjustments were made, and a variable M stores the number of adjustments.

While, alternative (1) consumes more memory, alternative (2) requires that an appropriate mapping be used to access pressure and temperature values at the same time instant. For example, at the *i*th time instant the values of pressures can be

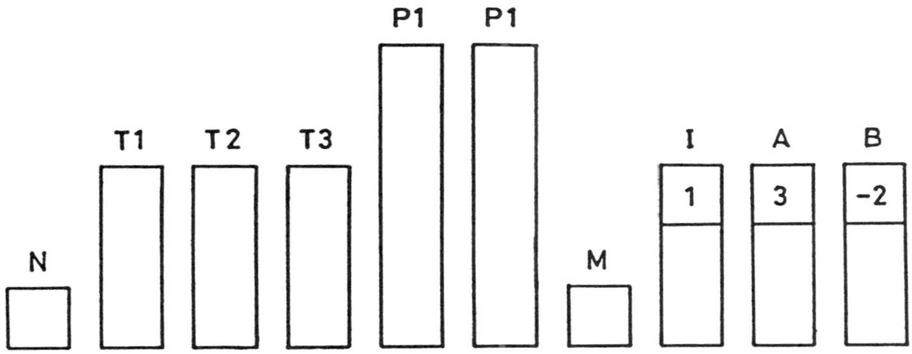

Figure 2.2—Data Structure with Five Data and Three Index Arrays

obtained by a lookup of P1(*i*) and P2(*i*), but the temperatures are obtained by a lookup of T1(*i*/3), T2(*i*/3) and T3(*i*/3). The valve adjustment at a given instant, however, can no longer be obtained by direct lookup. Instead, the array I needs to be searched for the value of the time instant. A saving in memory results in an increase in processing time.

Alternative (3) below lies between alternatives (1) and (2) as far as the memory usage is concerned, but it is as fast as (1) in the lookup of the adjustments, given a time instant.

3. There are five arrays for temperatures and pressures as before (Fig. 2.3), but a sixth pointer array (J) indexes into arrays A and B for the valve adjustments. Thus, at time instant *i* the valve adjustments are found from A(J(*i*)) and B(J(*i*)). A zero indicates that no adjustment was made.

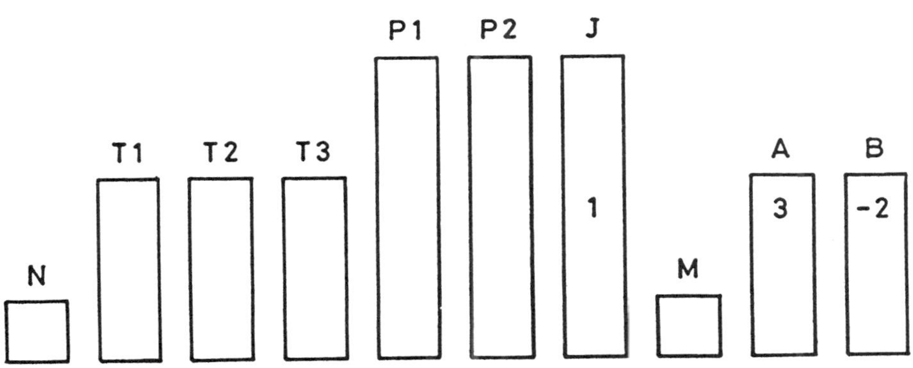

Figure 2.3—Data Structure with Five Data and One Pointer Arrays

When given an adjustment in this data structure, however, we cannot directly determine the time instant at which it was made. J has to be searched. If we wish to avoid a search, array I (as in alternative (2)) will have to be stored as well.

The above example brings home two important points:

1. There are space-time trade-offs in the choice of data structures. For instance, while alternative (2) consumes less space, alternative (3) is faster for lookup operations, given a time instant.

2. The choice of the data structures is made by anticipating or determining the kinds of uses or operations that are to be performed on it. For instance, alternative (3) should be selected if the lookup operation for adjustments is expected to occur frequently. On the other hand, if sufficient memory and storage space is available, alternative (1) may be preferred for its flexibility.

2.3 Program Design

After arriving at a data structure, we should proceed to develop the programs that operate on them. A number of principles ought to be kept in mind while writing programs. They follow from two basic ideas:

1. The program should be easy for the readers to read. It should be such that anyone (including the creator of the program) who reads it finds it easy to understand. Frequently, bugs will be found or changes will be made after the program has been written. To make the necessary modifications, somebody other than the creator may have to read and understand the program. This, as will be shown, is accomplished by following a certain discipline in writing programs and by providing adequate documentation.

2. The program should be easy for the writers to write. The program should be such that people who write it or update and modify it can do so easily. Frequently the application will undergo a change (e.g., an additional pressure sensor is added). The data structures and programs should be such that it is possible to make small changes in them to meet the new (slightly different) applications.

In the sections that follow, the design principles to achieve the above are explained.

2.3.1 Data Abstraction

If certain data structures are manipulated from many different places in the program, it is a good idea to define a set of routines (includes functions and subroutines) that perform the manipulation and to call those functions whenever the data structure is to be manipulated.

The first advantage of this is that it makes debugging easier while writing the program. Once the set of routines is defined and debugged, it assures that the lowest layer is working and protects the programmer from making mistakes. Second, if the data structure is changed later to improve efficiency or to reflect changes in the application, only the set specific to the functions needs to be changed. Calls to the routines do not need to be changed. This makes changes not only smaller in number but also limits them to a few routines.

For example, if we had a subroutine for accessing the values of the valve adjustments, then even if the data structure were to be changed from alternatives (1) to (2) or (3), the rest of the programs that use the valve adjustments remain unchanged. Only the subroutine has to be changed. If, on the other hand, arrays A and B were accessed directly, then a change in the data structure will make it necessary to painstakingly go through the entire program and make appropriate changes.

Similarly, for example, if we wish to store a new valve adjustment and we define a subroutine to do it, once it has been debugged it can be used with confidence. On the other hand, in its absence we will have to repeat the same thing whenever a new value is to be stored. (The latter case may become a source of errors. For example, we might at times miss updating M for the data structures in alternatives (2) and (3)).

2.3.2 Modularity

While developing software, it is extremely important that the problem be broken into subproblems and so on down to very simple tasks, and this structure should be reflected in the program. This can be done by writing subroutines for each of the subproblems and even separating or demarcating the data that pertains to each. For example, the control of the experiment described earlier might be subdivided as in Fig. 2.4.

Modular programs are easier to debug because the subroutines can be tested one at a time. They are also easier to understand because the problem structure is reflected in the program structure. A single module should not exceed one or two pages of code. Otherwise, it will be difficult to comprehend and, hence, be prone to errors. If a procedure is becoming very long, it is advisable to break it up into two or more shorter procedures.

When a program is being developed by a team of programmers, modularity at the level of subroutines is indispensable. The details of interfaces between subroutines

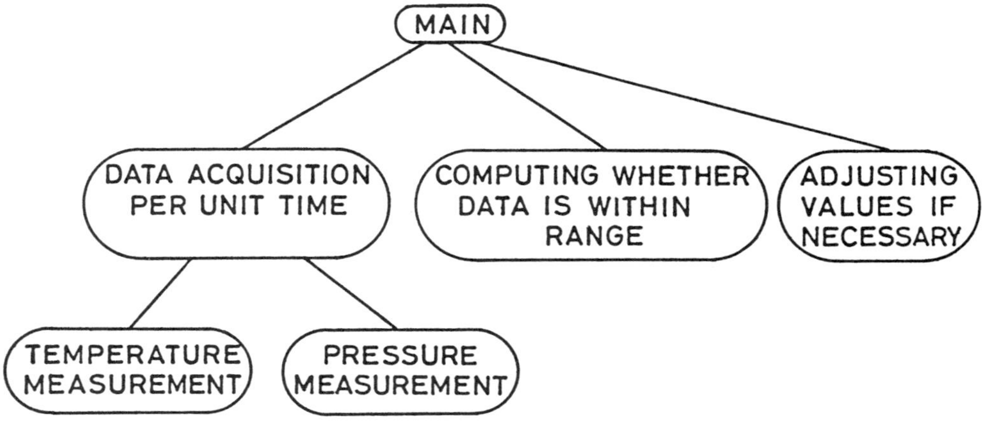

Figure 2.4—Modular Structure

have to be taken care of as well. Only then the work can be divided among the team members. Different subroutines can be parceled to different programmers. This obviously requires that the interfaces between subroutines be defined very clearly and precisely.

For example, if one programmer were to develop the data acquisition subroutines and another the range computation subroutines, there must be agreement on the common data structures, and the tasks to be performed by each must be clearly demarcated.

2.3.3 Structured Programming

It has been found that indiscriminate use of GOTOs makes the program extremely difficult to understand. Hence, their use in a program should follow a strict discipline as outlined below. Such a program is said to be structured. In general, every subroutine should have only one entry and one exit. Spaghetti or noodle programming should be avoided.

Several common but important structured statements are given below. If these are not available in the language(s) being used, they should be simulated. The structured statements follow from the PASCAL syntax, while the implementation language is taken to be BASIC.

IF-THEN-ELSE

The if-then-else statement in PASCAL is of the form

```
IF <cond> THEN <then-stmt-1> ELSE <else-stmt-1>
```

Many languages, including several implementations of BASIC, do not have an if-then-else statement; they have only an if-then statement.

```
IF <cond> <stmt>
```

In such a statement, if the IF <cond> is true, <stmt> is executed. The following simulates if-then-else.

```
          IF <cond>   GOTO  <n>
<m>       <else-stmt-2>
          GOTO <p>
<n>       <THEN-stmt-1>
<p>       <stmt-after-if then else>
```

This form of the use of GOTOs follows a rigid discipline and is acceptable.

WHILE-loop and UNTIL-loop

Most programmers may be familiar with the FOR loop in PASCAL or BASIC. It can be used when the number of iterations is known before its execution. There are times when such is not the case. Then after every iteration it must be checked whether the loop should be continued or exited. This is done by the while-loop and until-loop.

In PASCAL this has the form

```
WHILE <condition> DO <statements>
```

In Microsoft BASIC on the PC, the form is

```
WHILE <condition>
<statement>
:
:
<statement>
WEND
```

In versions of BASIC where WHILE constructs are not available:

```
<j>         IF <condition>  GOTO  <m>
            GOTO   <n>
<m>         <statement 1>
            .....
            <statement k>
            GOTO <j>
<n>         <statement-after-while-loop>
```

If the condition <cond> is true, the body of the loop is entered and the statements are executed. After that the test is repeated. If <cond> evaluates to false, the loop is exited and the control passes to the next statement.

REPEAT UNTIL statement

The form of the until-loop is:

```
REPEAT <stmt 1> ..... <stmt k> UNTIL <cond>
```

It differs from the while-loop in two respects. First, the iteration is carried out at least once before testing the condition. Second, if the condition is false, the loop continues rather than the other way around.

```
<j>         <stmt 1>
            ...
            <stmt k>
            IF <cond>  GOTO <n>
            GOTO <j>
<n>         <stmt-after-until-loop>
```

It is recommended that GOTOs should only be used as part of the above program forms. They should not be used for jumping around in the program indiscriminately.

2.4 Documentation

This is an extremely important aid to understanding a program. It is in the form of comments interspersed with the program code and in the form of manuals (separate or at the beginning of the program) describing the data structures and overall design decisions. It is strongly recommended that both these be present. An overall description of the data structures, frequently missed by beginners, is the most important reading aid.

For example, Routine 1 below might be the overall comments for the data acquisition and control example discussed earlier:

Routine 1

```
10   REM  This program acquires data from:
20   REM            three temperature sensors
30   REM            and two pressure sensors
40   REM  Data is stored in arrays T1, T2, T3, P1 and P2.
50   REM  T1(i), T2(i), T3(i)   -  temperature at time (i/3)
60   REM  P1(i), P2(i)  - pressure at time (i)
70   REM  J(i) = 0 if no adjustment is made at time (i)
80   REM         k>0,    where A(k), B(k) store the adjustment
90   REM  N    -   size of P1, P2, i.e., no. of pressure points
100  REM  M    -   size of A, B, i.e., no. of valve adjustments
```

It is a good practice to give comments at the beginning of every routine explaining:

1. what it does (in general terms);
2. inputs it takes (both as parameters and as global variables);
3. outputs it produces (value a function returns, for example); and
4. side-effects it has (global variables modified).

For example, the data acquisition subroutine may begin as shown in Routine 2 below:

Routine 2

```
1000 REM  DATA ACQUISITION SUBROUTINE
1010 REM  Dated : 30 October 1988
1020 REM
1030 REM  This subroutine is called at every time instant.
1040 REM  It reads pressure from sensors on ports E0h and E1h and
1050 REM  stores them in P1 and P2. Similar reads are performed
1060 REM  at every third instant for temperature and stored in
1070 REM  T1, T2 and T3 in kelvins.
1080 REM
1090 REM  Input:    N
1100 REM  Output:   P1(N), P2(N)
1110 REM  T1(N/3), T2(N/3), T3(N/3) if N is divided by 3
1120 REM  Increments: N
```

Comments should also explain, if necessary, the temporary or local data structures used in a subroutine and the algorithm used. Comments should be liberally sprinkled in the code wherever the programmer feels the logic is tricky or an important action is being performed.

The comments should explain the intentions of the action. For example, the following is not a proper comment:

```
100 N=N+1   '  Increment N
```

The proper form of the comment will be

```
100 N=N+1   '  Increment count of measurements
```

While the former just repeats the obvious, the latter explains the purpose.

2.5 Top-Down Design

In complicated programs it is good practice is to separately list the global variables that are used or modified, as well as other routines called by that particular routine. This is normally done immediately after describing the purpose of the routine, along with its version and date of last update. This makes future changes very easy to incorporate. Furthermore, when one needs to use some routine(s) in another application, the listing of variables and functions makes the task very easy. By including the date of update (and version), housekeeping is simplified by identifying variants of the same routine in the system.

References

Dahl, O.J.; Dijkstra, E.W.; and Houe, C.A.R. 1972. *Structured Programming*. New York: Academic Press.

Wirth, N. *Systematic Programming*. 1977. Englewood Cliffs, NJ: Prentice-Hall.

Yourdon, E., and Constantine, L. K. 1979. *Structured Design*. Englewood Cliffs, NJ: Prentice-Hall.

3
Software: Tools and Techniques

Sandeep Garg

3.1 Introduction

A wide assortment of software is available for the PC. Some software, such as the operating system and its utilities, is supplied with the machine; other software may be bought or developed by the user. There is a large selection of ready-made software suited to a variety of applications, as well as a choice of excellent development environments, should the user choose to develop his or her own. The choice between ready-made software or developing one's own depends on the nature of the requirement and on the availability of software suited to one's particular need. This chapter provides an overview of the types of software available for a typical computer system and a discussion of emerging techniques in software development.

3.2 Operating System

The operating system (OS) is a fundamental software unit that interacts closely with the computer system hardware and handles task and resource management. The OS, also the primary interface between the user and the system, handles the start-up and close-down of applications. With the increased variety of peripherals and computing power available at ever-lowering cost, the traditional role of the OS is being redefined and expanded. Operating systems now offer support for a selection of peripherals and have features such as built-in networking support. The essential functions of an OS are discussed in the following paragraphs.

3.2.1 Resource Management

The resources of a computer system—the CPU, memory, disk space, video display, and peripherals—have to be efficiently utilized and managed for executing programs. A single execution of a program is termed a process. In a multitasking environment, where many processes may be executing at a time, the system resources have to be distributed between the processes. In a single tasking system, the OS manages the memory and disk space for the process, handing over all other resources to the application.

3.2.2 Task Management

The OS handles the start-up and the close-down of applications and allocates resources to processes while they are running. In a multitasking environment, the OS switches between processes, allocating the CPU to each in an action called time-slicing. The OS also handles synchronization among processes. When responding to events such as interrupts, the OS switches the context of the CPU to the appropriate handler for the event.

3.2.3 Interface to the User

The OS provides the first layer of interface between the computer and the user. At the lowest level, the OS prompts the user to enter a command, interprets the command, and then executes it. The command the user gives may be internal to the OS, or it may start an application. The OS provides many commands and utilities for file manipulation, system configuration, and the management of system resources (memory, printers, ports, etc.).

3.2.4 Interface to Application Software

The OS provides an interface between the system hardware and applications. This interface is known as the application programmer's interface (API). The API enables hardware-dependent details to be isolated from the application software. It also increases the robustness of the application, as the underlying system hardware may be modified or upgraded without affecting the application. APIs are now being designed to conform to a standard common among OSs in order to facilitate the porting of one application to another OS with minimal changes.

3.2.5 Protection and Security

In a multitasking environment, the OS protects processes from interfering with each other. It restricts the access of a process to resources that belong to that process only. In the event of a process crash, the OS shuts the offending process down without letting it affect the other processes in the system. In a multi-user environment (where more than one user may use the system simultaneously), the OS restricts access by other users to objects such as files belonging to another user.

3.3 Advances in the Field of Operating Systems

The field of operating systems has undergone rapid advances in the past few years. These advances have come about in response to the need to make computers easier to use and the need to share data and resources among computers. In what is known as down-sizing, minicomputers and mainframes are now being replaced with networks of personal computers, and users are demanding the features that were traditionally available on such larger systems on PC-based operating systems. Also, the power to implement many features demanded by users is now available on the desktop, and this has led to OSs for the PC becoming increasingly sophisticated. At least some of the following features are likely to be found in PC OSs.

3.3.1 Graphical User Interfaces

The area of the PC user interface has undergone a revolution in recent years. In the pursuit to make PCs easier to use, traditional command line user interfaces have given way to graphical user interfaces (GUIs). With a GUI, the screen typically resembles a desktop, with applications running in individual windows that may be resized and moved around on the desktop. Most operations are accomplished by pointing and clicking with a mouse on parts of the window or by choosing from a menu. Because of their use of familiar metaphors as on a desktop, GUIs are more intuitive to use and have a quicker learning curve than the command line interfaces. For instance, an application for controlling an instrument from a PC may display the front panel of the instrument on the screen, with operations being carried out by clicking on familiar objects such as switches and knobs. Virtually every major OS has a GUI available for it, and the new generation of OSs is being designed to be graphical from the ground up.

3.3.2 Multitasking

In a multitasking environment, several processes execute simultaneously, either independently or in synchronization with each other. For control applications, multitasking is often desirable, as many input and output channels need to be simultaneously monitored, along with other programs, for performing calculations. Even for ordinary applications, the ability to simultaneously execute several programs and to share data among them leads to increased productivity. In order to be effective, multitasking requires a powerful CPU as well as large amounts of RAM and disk space. The current generation of PCs finally has the system resources available to support multitasking OSs.

3.3.3 Networking Support

The need to share data and resources among computers has led to networks becoming an integral part of most computer systems. This has led to a new category of operating systems known as network operating systems. The network OS typically executes either along with or on top of the main OS in order to manage the network and to provide the user and applications with a transparent interface to the network. Using the network OS, the data and resources of any computer on the network can appear as extensions of the data and resources of the local computer.

3.3.4 Support for Real-Time Applications

While running many control and communication applications, the system must respond to an input or event almost instantaneously. With conventional OSs, including multitasking OSs in which switching between tasks takes place regularly, the response time for an event cannot be guaranteed. A real-time operating system (RTOS) immediately responds to an input and can switch the processor context appropriately within a set time limit, usually of the order of milliseconds. RTOSs generally have a priority-based process switching mechanism and are more granular in form than conventional OSs. RTOS are available in many forms: as kernels that run under an existing OS (RT-kernel under DOS); as extensions to an existing OS (HP real-time extensions to HP-UX); or as completely new OSs (QNX).

3.4 OS Options

The new features being offered by OSs today place considerable demands on system resources. GUIs require a lot of computing power as well as fast graphics and I/O. Multitasking requires large amounts of RAM and disk space and advanced processors with the capability of handling multiple processes and protection in hardware. Networks also require large disk space and fast I/O. The horsepower for handling all of these, as far as the PC is concerned, came only with the introduction of the 80386 family of processors by Intel. Since then, as GUIs, networking, and multitasking OSs have caught on, technologies such as local-bus video, graphics accelerators, caching controllers, and network coprocessors have appeared. As the 486 and the Pentium, along with other microprocessors such as the Mips® 4000 and Digital's Alpha AXP appear in affordable personal systems, GUIs, multitasking, and networking are now essential features of a PC OS.

With the increase in power of personal computers has come the demand for more sophisticated operating systems that are capable of harnessing this power. This has led to a number of operating systems for the PC, ranging from the relatively simple to the quite sophisticated, each with its particular advantages and disadvantages. The following are major players in the PC OS arena.

3.4.1 MSDOS/PCDOS

This is the original operating system for the PC, and despite its limitations is still the most popular. Over the years it has undergone many transformations, and new versions keep coming fairly regularly. It has the advantages of being simple, familiar, and not very demanding on system resources. It also has the widest range of applications available. Its disadvantages are that it is neither multitasking nor multi-user, it cannot adequately harness the power of modern day systems, and it is not very robust when running wayward applications. The demise of DOS has been predicted for many years, but it has survived by adapting to the trends of technology. DOS now supports memory management for the 80386 (and higher) family of processors and many enhancements and utilities to meet the needs of the user of today. Additionally, DOS has the greatest range of device drivers and development environments available, which makes it attractive for use in developing control applications. If the need is for a simple, familiar OS that does not use a disproportionate amount of system resources, DOS continues to remain an attractive choice for the PC.

3.4.2 Windows

Windows began life as an add-on GUI for DOS. Because DOS was not powerful enough, Windows took over duties such as memory and peripheral management and used DOS simply as a file-handling system. Over time, Windows added on multitasking, an aesthetically pleasing user interface, and a powerful API to emerge as an OS in its own right. Currently, Windows still runs on DOS, but on the appropriate hardware (386 and above) it provides full multitasking and a distinctive user interface. Due to its success in the market, a flood of applications have been written for Windows, and most major DOS applications are now moving over to Windows. Because of Windows' powerful API and standard user interface guidelines, all Windows applications have a standard look and feel, making them much easier to use when compared to their DOS counterparts. However, Windows requires considerable system resources such as RAM and disk space in order to provide satisfactory performance. Also, Windows programs are more complex than their DOS counterparts and, therefore, are that much more difficult to develop.

Windows for Workgroups™ is a version of Windows with integrated networking support. It has facilities for sharing data across networks between applications running on different machines. Windows for Workgroups is one of the pioneers of a new category of software known as groupware; i.e., applications that intrinsically support the concept of users working together as groups.

The newest incarnation of Windows—Windows NT™—is a new operating system from the ground up. It is a sophisticated OS and supports multitasking, security, and the standard Windows User Interface and has the ability to utilize powerful PCs and their peripherals (such as CD ROMs). In addition, it has support for multiprocessing systems (systems in which more than one CPU is used to provide increased computing power). Windows NT targets the newest generation of PCs, which have very powerful CPUs (Intel Pentium, Digital Alpha AXP™), large amounts of RAM and disk space, and peripherals such as CD ROM drives.

Windows has now established itself as a dominant OS for the PC. Applications available for it continue to grow in number and are more powerful and easy to use because of the Windows environment. Writing programs to run under Windows is by no means an easy exercise, but powerful languages and development environments that make programming for Windows easier are becoming available.

3.4.3 Unix

Unix has been available for over two decades and, although having started life as an OS for minicomputers, is now available for PCs to supercomputers. This, in fact, is one of the attractions of Unix: it offers a standard environment across almost all platforms. It is a powerful OS, with support for multitasking and multiple users. Over the years, networking has been integrated into Unix, and some manufacturers have extended Unix to support multiprocessing on their machines. Unix requires a powerful machine in order to run and is fairly demanding on RAM and disk space. Also, it not very user friendly, having cryptic commands with multiple options. GUIs (such as OSF/Motif™ and Open Look™, based on the X-Windows windowing system) now available on Unix from most vendors provide an easier user interface. Several vendors have extended Unix, offering additional features such as enhanced security and real-time extensions. Unix is now undergoing a process of consolidation among manufacturers, and Solaris™ from Sun Microsystems® and Univel™ from network vendor Novell® are emerging as multiplatform Unix implementations on personal systems. Unix has also spawned the POSIX standard, which seeks to provide a common API across different OSs as an aid to developing applications that may be easily ported between OSs.

3.4.4 OS/2

OS/2™, developed and supported by IBM, is the dark horse of the OS race. It was launched in 1987 with great expectations and almost died when it failed to satisfy most of them. Meanwhile, the majority of the computing public switched to Windows. However, OS/2 has since matured into a stable OS that supports multitasking and security. It has a GUI called Presentation Manager™ that is well integrated with the base OS. One of the attractions of OS/2 is its facility for running multiple DOS and Windows sessions in a safe manner. Since OS/2 is the brainchild of IBM, it is targeted more toward the business community and, thus, currently has only a limited range of software for scientific applications. However, under the compatibility box, it is capable of running both DOS and Windows applications, albeit sometimes with reduced performance.

3.5 Popular Software Categories and Applications

The PC probably has a commercial software package for any application to which computers have been put. When it was launched by IBM in 1981, the PC was targeted at the business community, and most applications at the time were business applications. Many different types of packages are now available, but the typical applications for the PC in use in an R&D environment are the following.

3.5.1 Spreadsheets

A spreadsheet allows manipulation of numbers and their interrelations in tabular form, using a rectangular grid format. Each cell in the grid can be addressed, and values or formulae can be placed in it. Although a spreadsheet is not very useful in control applications, it is very useful in the analysis and plotting of data. Lotus 123 and MicroSoft Excel™ are the most popular spreadsheets available for the PC under DOS and Windows, respectively. Borland's Quattro™ runs a close second in both categories.

3.5.2 Databases

A database is an application for storing data in a structured manner. This structure is imposed by user-defined relationships between individual data items. Data can then be selectively retrieved by specifying these relationships. This is done either by using a menu-driven interface or by writing a program in a high-level language that is specific to the database. Structured query language (SQL, pronounced as Sequel), a method for accessing databases, is a standard language available across all platforms. For the PC, dBASE™ and Paradox™ from Borland are the database market leaders. Foxbase™ and Clipper™ are dBASE clones with enhanced performance and increased features. Access from MicroSoft is a popular Windows-based package, while Oracle™, Sybase™, and Ingres™ are very popular databases for larger systems. With the adoption of SQL becoming widespread, the actual database in use is becoming increasingly transparent to the user.

3.5.3 Wordprocessors

A wordprocessor constitutes an application for creating documents. Modern wordprocessors not only provide WYSIWYG (what you see is what you get) text entry and formatting facilities but can also automatically produce indices, table of contents, footnotes and endnotes, cross-references, and so on. They also include sophisticated spell checking and thesaurus utilities, and grammar and syntax checking is now being increasingly integrated. As more wordprocessing packages are being moved to Windows, the WYSIWYG and formatting capabilities are becoming more powerful, approaching those of desktop publishing packages. For the PC, WordPerfect, MicroSoft Word and Lotus Ami™ are the most popular wordprocessors, together commanding more than 95% of the market.

Until recently, scientific wordprocessors were a separate category, having extensive symbol- and table-handling capabilities. However, most of these capabilities have been progressively incorporated into the mainstream packages, and, as a result, scientific wordprocessors are being used for specialized tasks only. T_EX™, a very powerful text-formatting application, continues to be popular among the mathematical and physical sciences community, with many publications and professional societies accepting submissions in this format.

3.5.4 Presentation and Graphing Software

In R&D work, data often needs to be presented in the form of meaningful graphs. Spreadsheets provide limited capability in this regard. Presentation packages allow various graphical interpretations of data and can make different types of charts as well. These graphs and charts can be annotated with text and graphics and output to a printer, a plotter, or a slide maker. The leading packages in this field are Harvard Graphics™ and Lotus Freelance™ under DOS, and MicroSoft Powerpoint™ under Windows. Corel® Draw™ combines presentation graphics with drawing and painting capabilities into a very powerful and popular package.

3.5.5 Scientific Application Software

A category of software that is popular among scientists and engineers is the scientific scratchpad. These packages let the user enter values and formulae and then manipulate them at will. Graphing of variables and visualization of complex functions are major applications of these packages. The more powerful packages such as Mathematica™ and Macsyma™ have powerful symbolic manipulation capabilities as well. These packages, though expensive and slow, have placed immensely powerful mathematical capabilities in the hands of the researcher.

3.6 Programming Languages

For developing any application, the choice of programming language plays a key role. The options for programming languages available for the PC are discussed in detail in Chapter 4.

Programming languages can be divided into two categories: interpreted and compiled. Interpreted languages go through a program line by line, reading and executing each line in turn. Although this is a slow process, it is highly interactive and is therefore conducive to rapid program development. On the other hand, compiled languages go through a three-stage process: the entire program is compiled to machine code, then is linked together with libraries and other object modules to form an executable file, which is then executed. The execution speed obtained in the case of compiled programs is far higher than that of interpreted languages, and optimization of code for size or speed is possible. Modern integrated development environments (IDEs) integrate all the phases of program development, i.e., editing, compiling, linking, and executing into a user friendly environment. IDEs thus provide the convenience of an interpreted environment with the power of a compiled environment.

In addition to the compiler, programming languages have other utilities such as editors, debuggers, and libraries.

3.6.1 Editors

At its very simplest, an editor is a tool for entering and modifying code. Once word-processors did double duty as editors, but now powerful programmers' editors are available; editors used for program development now provide facilities for automatic indentation and keyword recognition, among other things, and can handle multiple windows and files. Almost all editors in IDEs have context specific help for the language.

3.6.2 Debuggers

A debugger (a tool for tracing program execution) is essential for developing error-free programs. Logical and run time errors in a program can be caught only with a debugger. The debugger executes the program in a controlled manner, allowing the user to single step through the code. The debugger lets the user watch variables and registers, manipulate these values, and set breakpoints and watchpoints, etc. Typically, for tracking down a bug, the user sets breakpoints and watchpoints around the suspect area of code. Then, by single stepping and monitoring values of variables, the bug can be tracked down, isolated, and rectified.

A beginner often tends to underestimate the importance of a debugger; suffice it to say that no serious developer would consider using a compiler unless a decent debugger is available for it. With the extreme flexibility of C, a debugger is an absolute necessity for any kind of development in the language. With IDEs, debuggers are tightly integrated with compilers and editors and are very easy to use.

3.6.3 Run Time Library

The run time library (RTL) is a library of routines supplied with the language. The purpose of these routines is to enhance the language and to provide the user with optimized code for most common programming tasks. Languages today provide increasingly sophisticated RTLs. Each language package may contain several versions of the RTL, tailored to different programming conditions. With the advent of object-oriented programming (OOPS, introduced later in this chapter), RTLs have become full-fledged application frameworks, which function as a template to which the user need only add the code specific to the task at hand in order to have a fully functional program that incorporates many features of commercial applications. However, because of their general-purpose nature, RTLs add overhead to the program, resulting in an increase in code size and a decrease in execution speed.

3.6.4 Third-Party Libraries

Third-party libraries bridge the gap between the needs of the user and the capabilities offered by compiler RTLs. Due to the variety of tasks to which the PC has been put, libraries of routines for a vast range of applications have been developed. These routines are written in a flexible way and can be easily incorporated into a program. Third-party libraries offer an excellent alternative to writing code for a number of applications. In the area of control and instrumentation, where system-dependent code needs to be developed, it is often more desirable to acquire a third-party library of routines than to develop one's own code.

3.7 Object-Oriented Programming

Object-oriented programming (OOP) is revolutionizing the way software is developed and maintained. Structured programming, the main programming paradigm practiced up to now, emphasized actions; OOP places equal emphasis on data and actions. Instead of programs being thought of as discrete pieces of software suited for a particular application, OOP promotes the concept of programs being reusable software components, which together can be used in flexible ways to solve a range of problems. This is a very attractive proposition for the developer, because code redundancy is cut down and programming effort saved by code reuse. This building block approach to software makes it easier to manage the complexity of large software programs, which are becoming increasingly common with the coming of GUIs, windowing systems, and multitasking.

Since OOP offers a new and different approach to program development, it is important to understand the basic OOP concepts. These are discussed in the following paragraphs.

3.7.1 Object

An object can be simply described as a piece of active data. Traditionally, data was a passive entity, being manipulated by procedures external to the data. An object contains both data and the code for accessing and operating on that data. The interfaces for accessing and operating on the data are well defined, and entities external to the object have restricted access to the components of an object.

Objects communicate by passing messages. A message is a request for action by an object; syntactically, it may look like a procedure call. A message from another object invokes procedures (called methods) within an object, which performs the operation and communicates the result to the requesting object. This mechanism separates code and data into logical function groups, with little chance of interference between groups. An entire program can then be divided into a number of interacting objects. Additionally, objects have properties (described below) that make them very powerful techniques for software development.

3.7.2 Class

Class is the definition of a group of objects that share common features and characteristics. These features and characteristics are defined in a base class; subsequent classes are defined as specializations of the base class. Classes and objects are analogous to data types and the actual instances of variables in structured programming. Objects are actual instances of a class.

3.7.3 Inheritance

Classes are usually arranged in a tree-structured hierarchy, with complex classes being derived from simpler classes by the progressive addition of more data and code. This property is known as inheritance, because classes inherit the behavior of other classes in the hierarchy. Classes that inherit properties from a base class may, in turn, serve as the base definition for other classes. Conversely, objects may be subclassed from a higher class to include only a subset of the properties of that class. Inheritance enables the progressive building up of objects suited to a particular task by deriving a new class from the appropriate classes.

3.7.4 Encapsulation

In traditional programming techniques, any data object may be modified by the program code. In OOP, encapsulation bundles data together with the code for the purpose of accessing and modifying the data within an object. The access to the components of an object can be controlled by the programmer. This way, an object can be made a relatively independent entity with a well-defined interface for operation. Since only procedures that are local to data can modify the data, any error in data values can be easily traced and bugs isolated in the code. This promotes the idea of a library of robust, well-designed objects that can be enhanced and reused.

3.7.5 Polymorphism

OOP often involves putting many different objects of unknown but similar types into a collection and then activating the objects in turn by sending each the same message. Each object responds to the message in its specific domain. This property of objects is called polymorphism. Polymorphism simplifies a programmer's task by generalizing the interface to objects while still retaining the internal distinction between objects.

Suggested Reading

Voss, G. 1991. *Object Oriented Programming: An Introduction*. Berkeley, CA: Osborne McGraw Hill.

4
Language Options

Sanjay Gupta

4.1 Introduction

This chapter deals with the various options available for the development of software and the criteria for the selection of a programming language for the purpose. The choice of software always contains a strong personal element. While, on one hand, suitability, convenience, cost, and so on are important parameters in its selection, an equally important factor is the personal preference of the user. Given two equally acceptable options, the programmer will naturally opt for the one with which he or she is more familiar. Software for data acquisition and control often has a large number of users and requires frequent changes. Such software tends to evolve over a period of time and with contributions from a succession of workers. This requires that the language in which it is written should be a popular one. The requirements of the particular application in terms of speed, graphics, etc., also affect the language selection. Quite often the end user may not be very conversant with computers. This puts additional demands of user friendliness, protection, and simple operation.

4.2 Compiler vs. Interpreter

Earlier machines tended to be slow and had poor operating environments. Under these conditions, the first choice facing the user was whether to use an interpreted language (usually BASIC) or a compiled language like C, Pascal, or FORTRAN. An interpreted language had an inherent appeal in that one could edit and run applications without any overheads. On the other hand, working with a compiled language, such as Pascal or FORTRAN, involved a long sequence of editing, compiling, and linking, which significantly slowed down the software development. Thus, for rough and ready work, an interpreted language had obvious advantages. On the other hand, a compiled language offered much higher speed.

However, with the advent of integrated development environments running on fast machines, the use of interpreters has rightly fallen out of use. Also, as users have realized the advantages of structured languages (which are almost invariably compiled), interpreted languages have fallen by the wayside. The subsequent sections present some popular languages and their merits and demerits for interfacing. The use of modular structured programming methods (OOPS) has grown very rapidly. OOPS was discussed in Chapter 3.

4.3 Some Popular Languages

4.3.1 BASIC

The first obvious option is to use BASIC (Beginners All-Purpose Symbolic Instruction Code), which comes with the operating system (OS), since it is already there and does not cost anything extra. It allows full access to the input/output and graphics capabilities of the computer, and software handlers are available for BASIC with most interface cards. In addition, compilers, available at a reasonable cost, can be used for compiling programs once they have been tested and debugged. Also, most of the older workers are familiar with FORTRAN, and BASIC is very similar. It is a simple language for the beginner to learn, having its origin as a simple interactive language for beginners. These are the reasons that a significant fraction of the work gets done in BASIC.

However, several drawbacks with BASIC render it less than ideal for the laboratory. First, the various versions of GWBASIC and BASICA that were shipped with versions of DOS, up to 4.xx were not structured, making modular programming difficult. However, from MSDOS 5.0 onwards, MicroSoft ships QuickBasic, which, while maintaining compatibility with the earlier versions (the files must be in ASCII, i.e., saved with the A option), allows structured programming. It should be kept in mind that BASICA (which was part of PCDOS) and GWBASIC (which came with MSDOS) can be considered as equivalent for almost all practical purposes. This often necessitated extensive rewriting of programs during development. Such occurrences with structured languages are far fewer. Furthermore, the handling of subroutines or procedures was virtually nonexistent in BASIC. This makes it rather difficult to develop standard procedures that can then be used in different programs. However, newer (structured) versions of BASIC (QuickBasic, Power Basic, Visual Basic, etc.) have overcome these limitations, and it is quite feasible to develop large applications under BASIC with ease. Many developers are switching to Visual Basic, especially under Windows, and application libraries in Visual Basic are becoming increasingly available.

The following is a summary of the advantages and disadvantages of BASIC.

Advantages

1. It is cheap (the interpreter is free with MSDOS).

2. It is easy to use.
3. It allows full access to graphics, etc.
4. It has good string-handling capabilities.
5. Good BASIC compilers are available.
6. Software handlers for BASIC are available for almost all interface cards.
7. Most versions have built-in interrupt handlers for the serial ports. Most other languages require supplementary drivers for this.

Disadvantages

1. Subroutines and procedures-handling are weak, making parameter passing rather difficult. Also, standardization is lacking. Thus, standard procedures cannot be developed.
2. It is not a structured language, and, hence, modular programming is not possible, making frequent rewriting of the software necessary. This problem has been overcome in the more recent versions.
3. It is a rather slow language.

Therefore, even though BASIC is a good language to use, it cannot be the only language for the R&D worker. Many interface cards provide the software interface to the BASIC interpreter free of cost or at a nominal charge. With BASIC compilers, a separate software interface is usually required. Otherwise, the programs require very few or no changes, at least in the initial stages.

4.3.2 Fortran

Most scientists who learned programming in the sixties and early seventies cut their teeth on FORTRAN and are thus bound to think in terms of FORTRAN as the language of choice in the laboratory. Furthermore, the volume of scientific software written in FORTRAN is extremely large. Unfortunately, FORTRAN is not very suitable for interfacing applications. Its string-handling capabilities are very poor, and it does not provide many of the other facilities such as port addressing, timing, etc., that are standard on modern languages such as C, Pascal, and Modula. Working with IEEE488, almost all instruments communicate with the computer through strings. Fairly extensive string-handling subroutines are required to generate commands, strip the additional information from the data, and convert the data to a numeric format. These problems have been addressed in the 1977 and 1990 FORTRAN standards, which has vastly improved its character-handling capabilities. With the passing of time FORTRAN is increasingly taking over many attributes of structured programming languages and is starting to look and feel more and more like Pascal.

FORTRAN compilers on the PC do not provide direct access to the graphics capabilities of the system in a standard fashion. This severely restricts the portability of software, which can be a major handicap. These failings are primarily due to the fact that, unlike BASIC, FORTRAN was never intended for an interactive environment; but for data analysis, there is probably more software available in FORTRAN than in any other language. Therefore, even though FORTRAN is not best suited for interfacing, sometimes it becomes necessary to use it. Most interface card suppliers do offer interfaces for the MicroSoft FORTRAN. Other versions of the language (Lahey, Salford, etc.) are also supported in a limited fashion. However, it is often possible to use .OBJ files from C with FORTRAN compilers. Mixed language programming is a specialized application and is avoided unless absolutely necessary.

The following is a summary of the advantages and disadvantages of FORTRAN.

Advantages

1. Most users are already familiar with FORTRAN.
2. Extensive software libraries are available for FORTRAN.
3. Libraries of commonly used subroutines and functions can be easily developed and used.
4. It permits access to routines written in machine language as well as to the input and output ports.
5. Most FORTRAN compilers are good, and programs execute fast.
6. FORTRAN compilers support the numeric coprocessor.

Disadvantages

1. The software development overheads are high.
2. Access to the PC's graphics capabilities is not available.
3. FORTRAN has very poor string-handling capabilities.

4.3.3 Pascal

Pascal does not suffer from the major failing of FORTRAN; it has satisfactory string-handling capabilities. However, some commonly used mathematical operations (e.g., exponentiation is a^b) are not readily available. It is also a structured language, which allows the easy development of modular software. Pascal was developed as a teaching language and, as is the case with most languages from the ALGOL stable, was weak in input/output capabilities. Borland expanded these capabilities with Turbo Pascal, which has emerged as the *de facto* standard, and almost all compilers now support this standard. Software handlers are available for Turbo Pascal, and most Pascal compilers also allow the use of MicroSoft-compatible .OBJ files from C.

However, Pascal is being supplanted by C, and the support for this language is progressively declining. Even though it can be argued that Pascal (and its sibling, MODULA) is probably the best language for applications, the trend towards C is almost irresistible. Now that Pascal has ceased to be a mainline programming language, support for it is rapidly decreasing. The latest version of Turbo Pascal from Borland, called the Borland Pascal, provides a flat memory model (making the segment–offset memory addressing on the Intel 80×86 family transparent to the user) coupled with the superb user interface and tools library, transparent (or nearly so) windows support, etc. With this development one may see a resurgence of Pascal for applications programming. However, with the rapid strides of Unix (and C) this may at best only temporarily slow down the apparently unstoppable takeover by C. The author has been observing the continuing decline of Pascal with regret. The advantages and disadvantages of Pascal can be summarized as follows.

Advantages

1. It is a structured language. Object-oriented programming is well supported in Pascal.

2. It has excellent string-handling capabilities.

3. Compiled programs execute fast.

Disadvantages

1. It has ceased to be a mainline development language.

2. Strong typing sometimes makes program coding somewhat complicated, as type mismatches have to be explicitly handled.

3. The input/output of Pascal has traditionally been weak. However, the *de facto* Turbo standard does not suffer from this.

4.3.4 C and C++

C is increasingly becoming the language of choice for system developers. Originally developed at the AT&T Bell Labs as a part of the Unix operating system, C has today become the standard tool for almost all software development. Developers have lately been switching to its object-oriented variant, C++™. This is of particular interest for interfacing applications since the modular nature of C++ code allows for rapid development of large applications by using a *mix and match* technique on previously developed modules.

With the increasing popularity of C, device drivers for it are almost invariably available. C is an immensely powerful language that permits the user to access almost every aspect of the system resources. This obviously reduces the level of protection available to the programmer against undesired operations that may be caused by programmer error. There are currently three accepted dialects of C: K&R™, ANSI, and C++. K&R conforms to the original specification of Kernighan & Richtie, ANSI to the subsequent ANSI standardization, and C++ the object-oriented variant developed by Stroustrup.

C language interfaces for device drivers are available in many cases. Some older drivers may not, however, support the ANSI variant. It should be borne in mind that C is primarily a language for the development of systems rather than for applications. While the low level of protection for the system from the user allows easy access to all its resources, it also requires much greater care in software development. On the other hand, the much greater portability of C programs may make it an ideal choice for prototyping and large control applications. Turbo PASCAL affords access to most system resources, with a higher level of protection, which is desirable for the average user. The extremely cryptic nature of C code also tends to make programs difficult to comprehend. Thus, while C may be preferred for prototyping applications, Turbo PASCAL may be preferable for routine interfacing applications. Since programs in C can run almost as fast as those in assembler, it may used in place of assembler to write fast timing critical code (as discussed in 4.5).

4.3.5 FORTH

FORTH proponents consider it to be the natural language for process control. However, very few manufacturers of interface cards for the PC list interfaces for FORTH in their catalogs. Thus, unless willing to develop his or her own device drivers or at least software interface to the driver, the user is out of luck. FORTH users may consider using a package such as ASYST™ (which is very much like FORTH) and for which device drivers are available. Also, many commonly used procedures are built in into these.

4.4 Special-Purpose Packages

A number of available special-purpose packages have been designed especially for laboratory data acquisition and analysis. These should not be confused with packages such as LABPAC (Scientific Solutions® or Tecmar®), PCLAB (Data Translation®), and GPIB-PC (National Instruments®), which are essentially software handlers for linking the card and the high-level languages. They are either machine language routines, libraries, or object code modules supplied by the manufacturer of the interface card.

Special-purpose packages are essentially full working environments that can be used with a number of interface cards for data acquisition, control, and analysis. Some examples are LABTECH NOTEBOOK™ (Laboratory Technologies®), ASYST™ (Macmillan®), DADiSP™ (Adept Scientific®), and ILS™ (Signal Technology®). These packages are not from manufacturers of a particular interface card but from third-party sources. Their prices and capabilities vary. The more popular interface cards are almost invariably supported. Rarer cards are not very likely to be supported by a package. These packages can be very useful for situations in which intensive use with frequent modifications is planned. It should be borne in mind that in case the package cannot do what the user wants, he or she will be forced to fall back on a high-level language.

4.5 Assembly Language

Almost every supplier of PC interfaces offers one for assembly language. Assembly language is the most powerful and also the most painful to use. The user can get by quite happily without ever having to program in assembler. It is much more convenient to let the compiler take care of converting one's thoughts to machine language through a high-level language rather than try to program in the machine language oneself.

However, in some situations it becomes necessary to make judicious use of the machine language. For applications in which speed is absolutely necessary, programming in assembly language may be the only way. On close analysis, even in these applications, it will be found that only a very small part of the program is critical in timing, and it is usually possible to write the rest of the program in a high-level language and include a machine language routine for the critical parts only. Typically in any program, 90% of the time is spent in executing 10% of the code. Judicious use of assembly language to code this 10% will often be all that is required to attain the required speed. Modern profilers make the task of identifying this part of the code relatively simple.

Quite often, machine language programming may appear to be necessary at first sight, but this is not always strictly so. In many such situations data transfer occurs in a "bursting" fashion; i.e., periods of intense data transfer are followed by relatively long periods of inactivity. In such cases, if the interface cards support DMA (direct memory access), data transfer can be done through the DMA mode at full speed through a high-level language routine. However, if the interface card does not support DMA, the data transfer routines may have to be coded in the machine language. In situations like this it is often more convenient to work in FORTRAN, Pascal, or C and link in these assembly language routines.

However, in situations in which both input and output must take place very fast, it may become necessary to program in the machine language. It may be kept in mind that in most cases the penalties of coding in a high-level language with an optimizing compiler may be minimal. For example, with C the speed penalty will generally be less than ten percent.

Thus, it can be safely said that, given the hardware and software currently available, it is very rare that the need to program in the machine language will arise. In most such situations, it is usually sufficient to simply write that part of the program in which timing is critical in the machine language and use a high-level language for the rest.

4.6 Conclusion

There is no perfect language for interfacing, just as there is no ideal language for all programming needs. In the author's experience, Turbo PASCAL comes very close to this elusive goal as far as interfacing is concerned. However, with PASCAL progressively ceasing to be a main line programming language, the author has switched to C (and C++), albeit with regret.

At one time the languages of choice in the author's laboratory were interpreted BASIC and Turbo PASCAL. However, now almost all work is done in C, even though the absence of rigid typing in C makes debugging somewhat more difficult. ANSI C is preferred for its faster execution as well as more rigid typing. C++ and OOPS are increasingly being adopted.

Suggested Reading

The reader is referred to the manuals of the various languages. There are often significant differences between versions of the same language, and the importance of using the correct manuals cannot be emphasized. Also, books on almost every language are available from various authors. For reviews of the various languages and packages, the reader may refer to journals such as *BYTE*, *PC World*, and *PC Magazine*.

Brown, D. L. 1985. *From Pascal to C*. Belmont, CA: Wadsworth Publishing Co.

Duncan, R. 1986. *Advanced MSDOS*. Redmond, WA: Microsoft Press.

Edwards, C. C. 1988. *Advanced Techniques in Turbo Pascal*. New Delhi: BPB Publications (under license from Sybex).

Jamsa, K. 1988. *Power User's Guide*. Berkeley, CA: Osborne-McGraw Hill.

Kernighan, B. W., and Richtie, D. M. 1988. *The C Programming Language (ANSI-C Version)*, 2nd Ed. Englewood Cliffs, NJ: Prentice Hall.

Schildt, H. 1990. *Using Turbo C++*. Berkeley, CA: Osborne-McGraw Hill.

Stroustrup, B. 1987. *The C++ Programming Language*. Reading, MA: Addison-Wesley.

5
Sensors and Associated Systems

S. S. Prabhu

5.1 Introduction

The past 30 years have witnessed explosive growth in information technology, which was ushered in by developments in solid-state electronics. Miniaturization and large scale integration have resulted in developments in analog and digital electronics that have enabled designers of complex systems, such as those in the process industry, to address themselves to tasks that would have seemed impossible not long ago. The present day computer revolution and the ability to build inexpensive microprocessor-based computer systems for industrial instrumentation and control have brought about fundamental changes in the way things are done in the industry. Thirty years ago, system integration was a distant dream, but it is now a fact of industrial life.

Microprocessors have become the significant element in smart sensors and smart actuators, programmable logic controllers, mini- and microcomputers, local area networks, robots, vision systems, etc. They have enabled system integration on a large scale in the process industry. Measurement, computation, communication, and control are of fundamental importance for this activity.

This chapter gives a general description of sensor systems and how to protect them from unwanted interfering voltages and currents. It also touches upon some recent sensor developments.

5.2 Frequently Measured Important Quantities

A list of some of the important quantities measured frequently in the process industry is given below. The list is not exhaustive and refers only to quantities that are often measured electronically. However, it gives an idea of the range of the diverse quantities measured in practice.

1. Temperature
2. Pressure (including vacuum)
3. Fluid flow

4. Level of liquids
5. Fluid density and specific gravity
6. Viscosity and consistency
7. pH
8. Humidity of gases
9. Moisture content in solids
10. Thermal conductivity
11. Calorific value
12. Force, tension, compression
13. Level of solids
14. Mass and mass flow rate
15. Thickness and displacement
16. Angular and linear velocity
17. Electrical conductivity
18. Optical properties, such as surface reflectivity, transmittance, refractivity
19. Electrical quantities such as resistance, inductance, capacitance, voltage, current, frequency, power, energy.
20. Composition by chemical analysis and spectroscopy (including infrared absorption, atomic absorption/emission, ultraviolet and X-ray absorption, mass spectroscopy, microwave spectroscopy, nuclear magnetic resonance).
21. Radioactivity

5.3 Sensor Systems

Sensor systems convert the physical quantity under measurement into a coded electrical signal, i.e., the value of the measured variable is coded, in some way, in the electrical signal. For example, the value of the signal may be coded in the form of an amplitude or frequency-modulated signal. Another example is a shaft encoder, which converts angular information into a Gray-coded binary digital output.

Sensors are generally associated with electronic systems for the purpose of signal conditioning and processing, multiplexing, conversion of signal from one form to another, etc. A block diagram of such a system is shown in Fig. 5.1. The functions of the various blocks in this figure will be briefly explained.

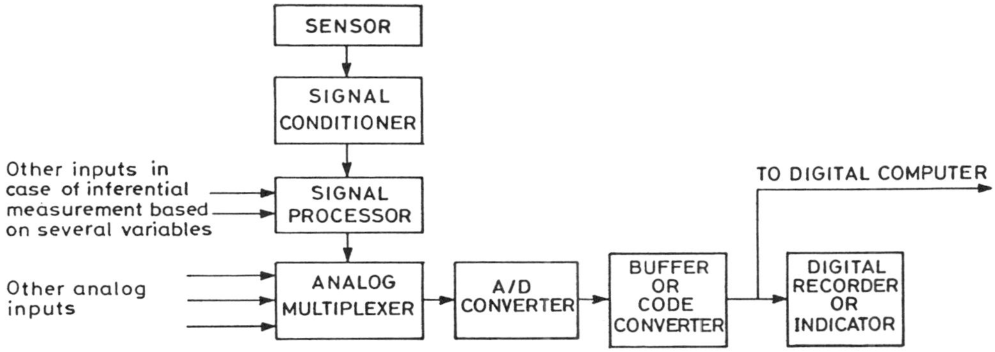

Figure 5.1—Typical Sensor System

The signal conditioner amplifies and suitably modifies the signal from the sensor. It generally includes the supporting circuitry of the sensor. Thus, it provides sensor excitation power, balancing circuits, and calibration elements. Signal conditioning has been defined to accomplish one or more of the following [3]:

1. Convert the sensor output into an electrical signal in a suitably coded form.
2. Using the outputs of two or more sensors, make an inferential measurement by computation according to a specified mathematical relationship and output a signal carrying this information.
3. Do linear or nonlinear amplification.
4. Filter unwanted frequencies from sensors and their associated circuits.
5. Compensate for limitations of sensors and/or receiving devices (for example, recorders) in order to extend their useful frequency range.
6. Provide impedance matching.
7. Compensate for thermoelectric errors.

As is evident from the above, signal conditioning is understood to also include a certain amount of signal processing. However, in general, signal processing operation is separate and distinct from the operation of signal conditioning. With advances in electronics, signal processing of a complex nature can now be readily accomplished, especially if microprocessors are used.

A large variety and range of signal conditioning modules for many types of transducers are commercially available. For example, thermocouple signal conditioning modules have the thermocouple voltage as input and provide reference junction compensation, amplification, and linearization. The output is often a standard ±5 V dc full-scale signal, which varies linearly with temperature. Signal conditioning modules generally have two outputs. One, which may be called the main output, is continuously available. The other, which may be called output, is for transmission (for example, onto the signal bus) and is available only when the module receives a command in the form of a logic level. This is shown in Fig 5.2. Thus, the thermocouple conditioner always gives an analog output proportional to the temperature and may also give a digital signal when required to do so.

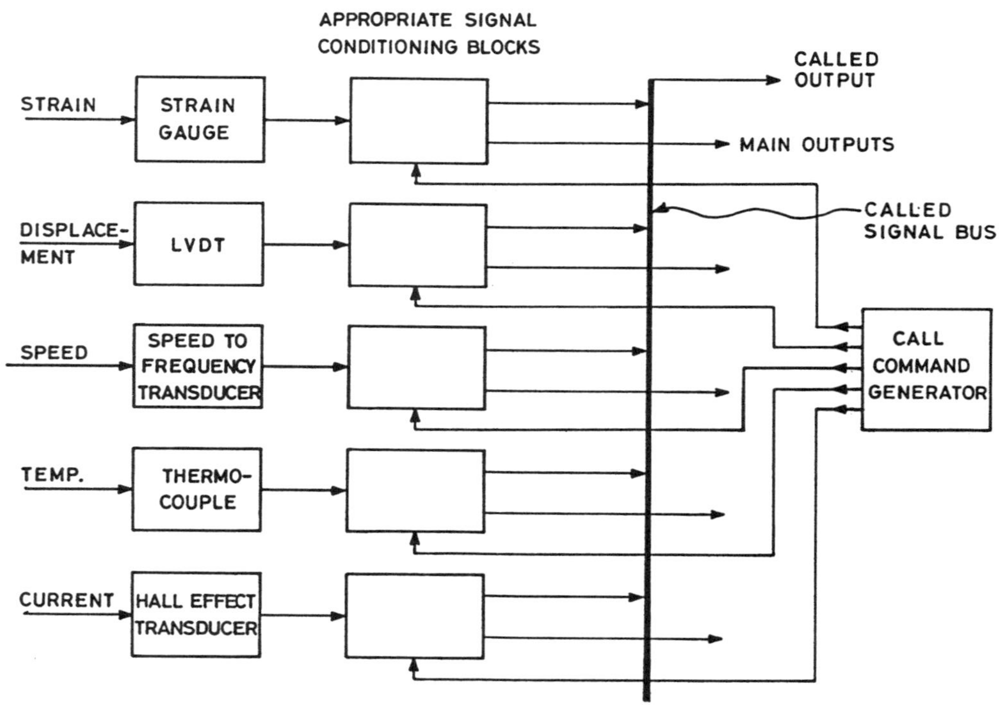

Figure 5.2—Signal Conditioning Modules in Operation

Many types of analog signal processing modules are commercially available. A certain amount of logical decision-making capability is generally incorporated in them. They perform such functions as multiplication of two or more input signals, square, square-root and absolute value evaluation, division of one signal by another, integration, limiting value monitoring, comparing two signals, and so on.

The analog multiplexer block shown in Fig 5.1 accepts several analog inputs and sequentially outputs one input at a time. Digital multiplexer inputs and outputs are in digital form. An A/D converter converts analog data to a digital form. The next block is either a simple buffer or a code converter to suitably code the digital data coming out of the A/D converter. For example, this may represent conversion of output bits emerging in parallel from the A/D converter to a serial form for transmission over a cable. The end device is a recorder or an indicator.

If transmission of a sensor signal to a distant location is needed, it becomes necessary to convert the signal to a suitable form that depends upon the nature of the telemetry link used. Long cables are susceptible to noise, voltage drops, and other problems. Thus, digital transmission of signals is preferred over long distances. This is shown in Fig. 5.3. For example, the incoming signal may be converted by the signal converter into a frequency-modulated signal to transmit analog data or into frequency shift keying and pulse code modulation to transmit digital data.

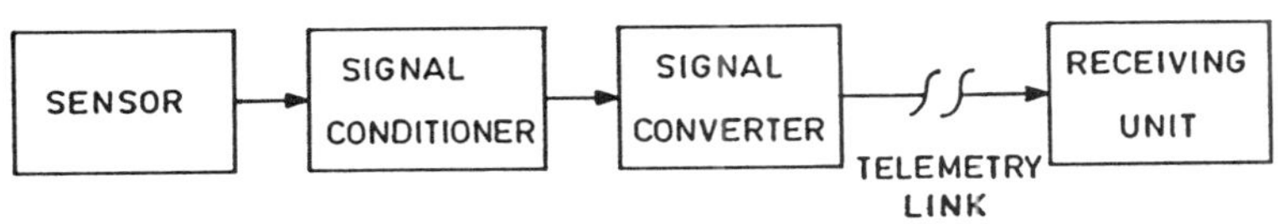

Figure 5.3—Signal Path in Long Distance Communication

Frequently, the relationship between the measured quantity and the sensor output is nonlinear, e.g., the output of a thermocouple. The nonlinearity can be canceled and a linear relationship obtained by electronic means. In situations where sensors are "smart," that is, when the sensor data is processed by a microprocessor system before transmission to the receiving unit, the linearization operation is conveniently performed digitally by the microprocessor.

5.4 Types of Data

Data from sensors can be broadly classified as static, dynamic, and transient. Static data are not subject to rapid fluctuations. They vary so slowly that they may be considered to be constant, e.g., the level of water in a large reservoir. Dynamic data, in contrast, represent rapidly varying quantities, for example, aircraft wing flutter. Transient data refer to the measurement of transient phenomena that manifest only for a specific, short time and then cease to exist. Surge voltage due to lightning is an example of a transient phenomenon. Analog measurements are usually more suited for dynamic data. If the rapidity of fluctuations of the quantity under measurement is high, i.e., if the signal bandwidth is large, analog processing of the sensor signal may be the only option available, and digital processing, with the available hardware, may not be feasible. For slowly varying data, digital techniques are highly suited. Furthermore, multiplexing of a large number of signals can be easily accomplished, resulting in economy in cabling costs among other things. Transients can be as slow as a few seconds or as fast as a few nanoseconds. Depending on its speed, analog or digital measurement can be employed. In any case, whenever sampling of a signal is done, care should be taken to see that the requirements of the Sampling Theorem [7] are met. In practice, it is advisable for the sampling frequency to be at least 4 to 5 times the signal bandwidth.

5.5 Types of Sensors

Sensors usually consist of an actuating mechanism and a sensing element. Some sensors may use the sensing element directly, such as the thermocouple and the photovoltaic cell. In others, some kind of actuating mechanism is used, such as bellows, Bourdon tubes, diaphragms, seismic mass, and spring, which give input to the sensor. Thus, in a pressure-sensing system, the actuating mechanism may be a diaphragm subjected to the pressure to be measured, and the sensor may be a strain gauge bridge fixed on the diaphragm.

Sensing elements can be classified in various ways. The classification may be based on the measured quantity, for example, force, torque, displacement, velocity, acceleration, pressure, volume and mass flow rate, and temperature. The classification may depend on the physical effects used in translating the input into output, for example, photoelectric, thermoelectric, electrochemical, and thermoresistive. Sensing elements may be active or passive.

Following Cerni and Foster [3], one can broadly classify sensing elements into the following categories:

1. Self-generating analog
2. Variable parameter analog
3. Frequency or pulse generating
4. Digital

Most sensing systems employ sensing elements of the analog form. Some examples of the above four categories are given below.

Self-Generating Analog: Piezoelectric, photo-electric, thermoelectric, electrodynamic, Hall effect.

Variable Parameter Analog: Variable resistance sensing elements, (for example, strain gauges, potentiometers, thermal resistance detectors, hot wire anemometers, semiconductor strain gauges, semiconductor resistance thermometers [thermistors, photoresistive]), variable inductance, variable capacitance, variable reluctance, variable permeability, magnetostrictive, variable mutual inductance sensors.

Frequency Generating: Fig. 5.4 gives an example of a pressure measuring system in which the output is a frequency signal, that is, a signal whose frequency is proportional to the measured pressure.

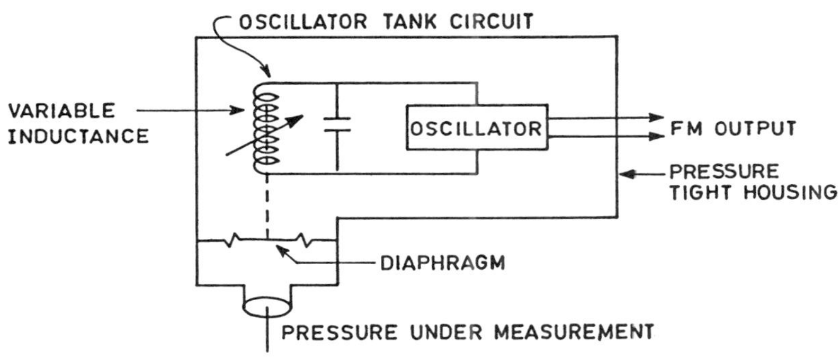

Figure 5.4—Pressure Sensor with Frequency Modulated Output

Pulse Generating: Fig. 5.5 gives an example of a position-measuring system in which the position information is given by the contents of a counter. Radioactivity sensors like Geiger-Muller tubes, scintillation counters, etc., are of the pulse generating kind.

Digital: Shaft encoder, in which the input angle information is quantized and converted into a digital output.

5.6 Amplification and Signal Processing

The electric signals produced by sensing elements are mostly at low voltage and low power levels, necessitating their amplification. The nature of the amplifier depends on the nature of the signal to be amplified. Thus, in situations where the signal is in

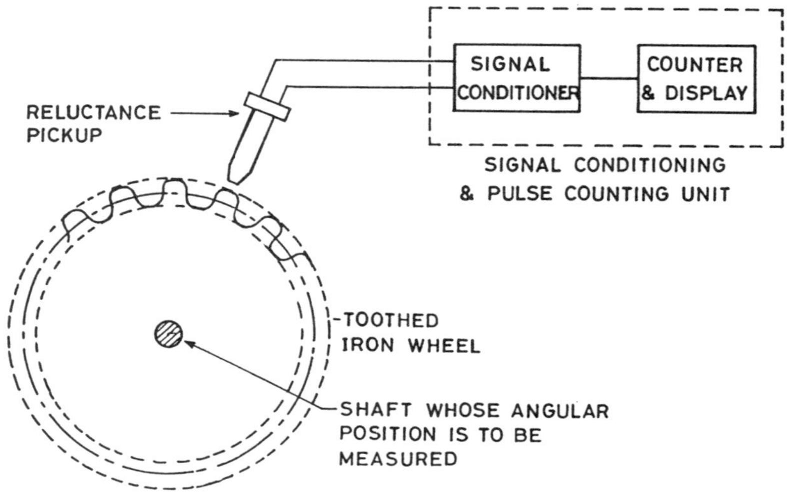

Figure 5.5—Counter-based Position Sensing System

amplitude or frequency-modulated form, appropriately tuned amplifiers have to be used. The most common situation for signal processing in practice is where arithmetic operations on analog signals are to be done. Op amp circuits are highly suited for this purpose and are widely used. Such operations as addition, subtraction, multiplication, division, integration, differentiation, function generation, square- root evaluation, and linearization can be readily done using these circuits. Op amps are also used for building active analog filters and various kinds of analog and digital signal processing equipment. There are many other kinds of basic operations, such as modulation and demodulation, analog-to-digital convertor (ADC) and digital-to-analog convertor (DAC) operations, etc. Special-purpose ICs are available for most such operations. Figure 5.6 gives an example of an analog instrumentation system in which some arithmetic operations are done with a view to obtaining signals proportional to the speed, power, and torque of a gas turbine.

5.6.1 Instrumentation Amplifiers

From the foregoing it is clear that many kinds of amplifiers are used in practice for signal conditioning and processing. However, a particular type of amplifier has been traditionally given the name instrumentation amplifier. Commercially, various types of such amplifiers are available. The major characteristics of instrumentation amplifiers are high common mode rejection ratio (typically 80 dB), high input impedance (1 megohm or higher), low noise (under 10 μV) and drift voltages (under 10 μV/deg)

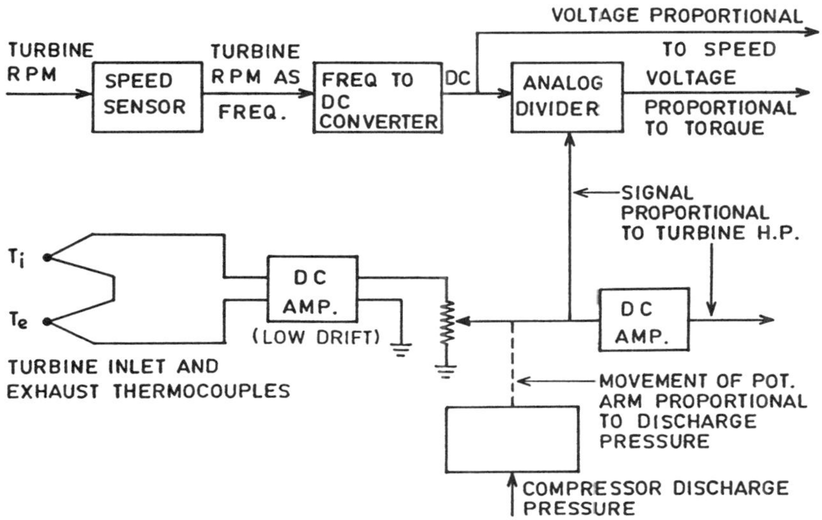

Figure 5.6—Gas Turbine Sensing Using Analog Signal Processing

in the output, a moderate band width, and a variable gain (usually 1 to 1000), which can be set by a single resistance manually or be programmed from a remote location by a microprocessor. A common mode rejection ratio of 80 dB and an input impedance of 108 ohms are readily available. Figure 5.7 shows a basic instrumentation amplifier configuration using three op amps.

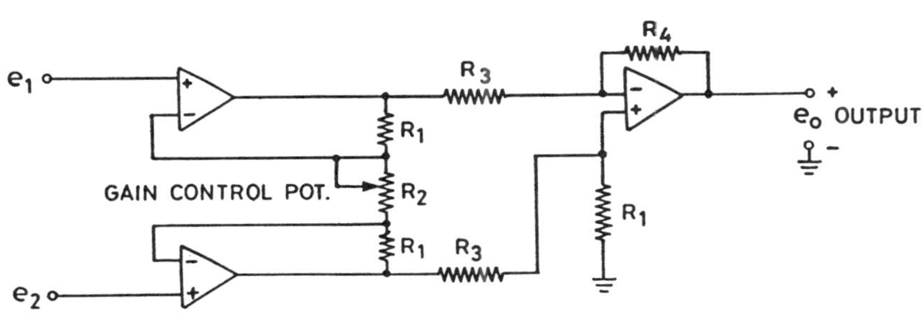

Figure 5.7—Instrumentation Amplifier Configured with Three Op Amps

5.6.2 Isolation Amplifiers

Isolation amplifiers are special instrumentation amplifiers for dealing with low level signals riding on high common mode voltages. They work in difficult electrical environments where there may exist ground disturbances, power transients, interference from electrical apparatus such as motors, switchgear, and power lines. These amplifiers do not require signal ground connection, that is, the signal input is isolated from ground and rejection of the ground disturbances is improved. The signal input circuit is electrically isolated from signal output as well as from the power supply, and in some cases there is also isolation between amplifier output and power supply. Such amplifiers are essential in certain medical electronic equipment for the safety of the patient. Isolation is achieved by transformer coupling and/or optical coupling. If transformer coupling is used, it is the usual practice to use signal modulation, ac amplification with transformer coupling, and demodulation. Opto-isolators convert electrical signals to light signals and back and in the process achieve isolation. Ordinary instrumentation amplifiers have typical common mode and differential voltage ranges of ±10 V, whereas, for isolation amplifiers these ranges may be as high as ±2500 V and ±250 V, respectively. References 9 and 10 give details of operation and comparison of instrumentation and isolation amplifiers.

5.7 Parasitic and Noise Voltages in Instrumentation Circuits [4, 5, 6]

In high accuracy measurement systems it is necessary to take special precautions to reduce noise voltages. One of the main sources of noise voltages is the utility power line. These voltages enter the system through coupling provided by parasitic or stray circuit parameters such as parasitic capacitances. Other types of noise are ripples due to an inadequately filtered dc supply, switching and line transients, noise due to pickup from electromagnetic radiation, thermal noise in electronic devices, noise due to ground currents, and so on. Proper shielding and grounding of sensitive portions of electronic circuits are used to avoid noise contamination. Due to the rapid degradation of the electrical environment of sensor systems in industrial situations, noise pollution has become a serious problem.

Apart from noise, there is also the problem of drift in the bias currents and offset voltages in amplifiers caused by variations in temperature, line voltage, and circuit parameters. Drifts and their effects have to be reduced in instrumentation amplifiers and circuits. Precautions must be taken to guard especially against the formation of unintentional thermojunctions in these circuits, thus introducing temperature-related drifts. If dc response is not required, ac coupling should be resorted to since it greatly reduces drift effects. In digital systems, the auto-zero technique can be used for drift compensation. Since sampling is involved in such systems, the input can be periodically shorted, and the resulting output, which is clearly due to drift effects only, measured and put in memory. When the next signal measurement is done, the drift voltage measured earlier can be subtracted to get the true value of the signal.

Noise can enter instrumentation circuits through inductive or capacitive coupling, ground loops, and even radiation coupling. The signal path may be protected from contaminating voltages by the use of a shield. Shielding is for the purpose of preventing either electric or magnetic coupling with the noise sources. Since contamination by electric coupling is more common, shielding against electrically coupled voltages is done quite frequently. A metal housing of the sensitive parts and proper grounding of the housing will prevent electric coupling. This may be accomplished by a housing of aluminum foil or, as in the case of signal cables, by a covering of braided copper sheath. Magnetic coupling is avoided by using shields of ferromagnetic materials with high relative permeability, in the range 10,000 to 100,000. Shields of nickel-iron alloys are common. In addition, care has to be taken to reduce inductive coupling by arranging the circuits in such a way that they do not, as far as possible, have loops into which voltages can be induced by changing the magnetic fields that link them; thus, two signal-carrying conductors should be twisted together to eliminate inductive coupling.

It is important to note that in electrical shielding it is possible to cause greater noise in the signal by improper grounding of the shield than would be present without shielding. It is, therefore, necessary to discuss the problem of grounding.

5.8 Grounding Techniques [4, 5, 6]

5.8.1 Safety Ground and Signal Ground

Ground design is an important and major task in the design of signal circuits. A brief discussion of grounding techniques follows.

There are two types of ground in electronic circuits, namely, safety or earth ground and signal ground.

All the external parts of instruments are connected to a point called the safety ground terminal and this, in turn, is connected to the ground lead of the ac utility power line cable. In a proper installation, this ground connection ensures that none of the exposed conducting parts can have a potential harmful to persons who come in contact with them. For ensuring this, the ground lead of the ac power line has to be properly earthed at some point, usually the switching or transformer station where the power line originates.

The signal ground in an electronic circuit is the point to which all the biasing and signal voltages are related. Signal grounds can be metal chassis, conduits, raceways, neutrals, racks in electrical equipment, or a properly earthed point. It should be noted that no two ground points can be truly at the same potential. This is especially so if large, unwanted currents flow through ground systems due to power line leakages or large unbalanced three-phase loads, or if there exist high radio frequency radiation-induced currents, as in the vicinity of transmitters.

5.8.2 The Shield Potential Problem

Consider Fig. 5.8a, which shows an ungrounded shield around an amplifier. C_1, C_2, and C_3 represent capacitances from shield to the amplifier input and output and signal ground, respectively. The equivalent circuit of the setup is shown in Fig. 5.8b. It is clear that there is feedback from output to input because of C_1, C_2, and C_3, which may cause instability. To avoid this problem, the capacitance C_3 is usually shorted by grounding the shield to the signal ground, as shown in Fig. 5.9a. Figure 5.9b is the equivalent circuit of the arrangement. It shows that now there is no coupling between output and input of the amplifier through capacitances to the shield.

Consider another problem of the floating (ungrounded) shield as shown in Figs. 5.10a and 5.10b. Here, the signal ground of the amplifier is brought out and connected to an external structure (ground). Parasitic capacitances can couple unwanted signals directly into the circuit. Again, if capacitance C_2 is shorted by connecting the shield to the signal ground, this problem is solved (Fig. 5.10b).

There are several choices for the location of this shield-to-reference conductor connection. Consider the circuit of Fig. 5.11. Here, the shield is connected to the signal ground of the amplifier at (3) and (4). Conductor (5) represents the return or low side of the low-level signal input to the amplifier; (6) represents the ground with respect to which the signal is defined. It is clear that the noise voltage source V causes current to flow through the return conductor (5). As a result of the resistance of (5), there will be a voltage drop on (5) due to noise source induced current. This causes a noise voltage to appear as an input to the amplifier. If this situation is to be avoided, the shield must be grounded in such a way that noise currents will not flow through the signal conductors. Figure 5.12 shows how this can be done in the case of the amplifier under consideration.

The following two rules emerge from the above discussions:

1. The shield must be connected to a zero-signal reference potential to avoid unwanted coupling.
2. The shield connection should be such that it prevents parasitic current flow through the signal conductors.

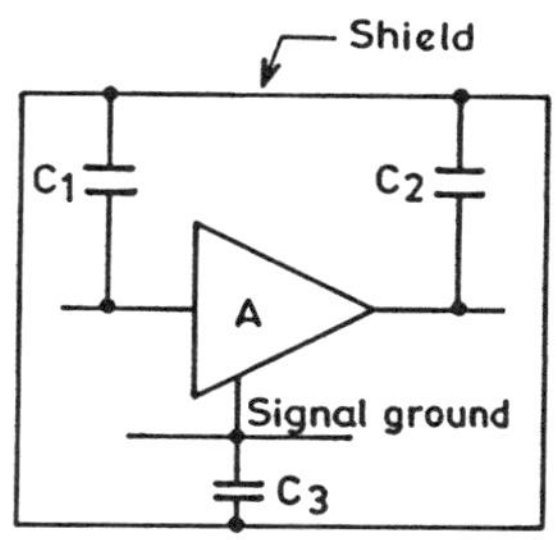

Figure 5.8a—Ungrounded Shield Around an Amplifier

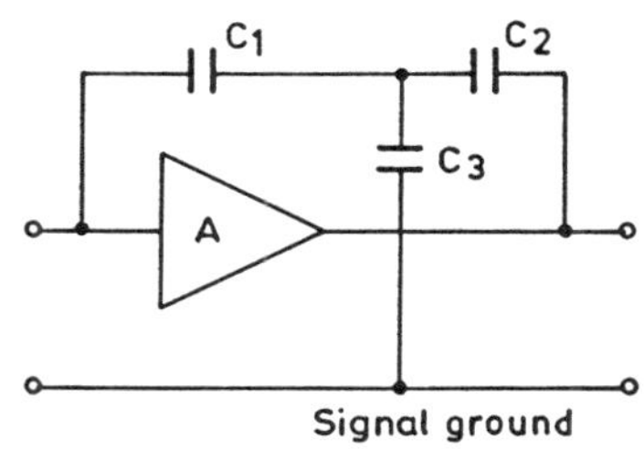

Figure 5.8b—Equivalent Capacitances for Fig. 5.8a

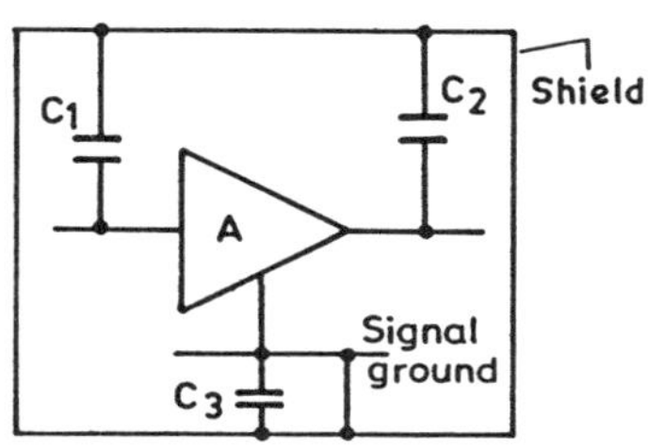

Figure 5.9a—Grounded Shield Around an Amplifier

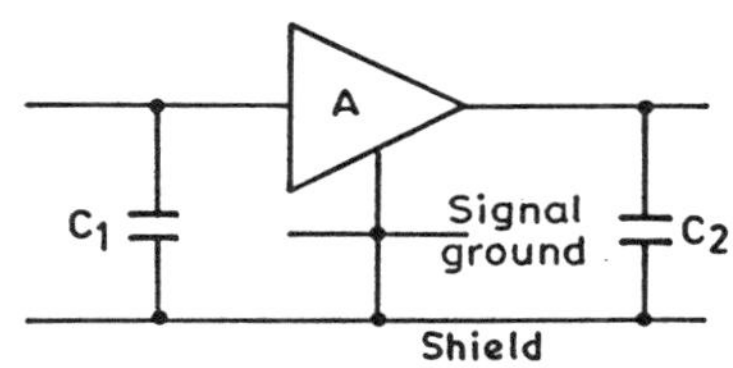

Figure 5.9b—Equivalent Capacitances for Fig. 5.9a

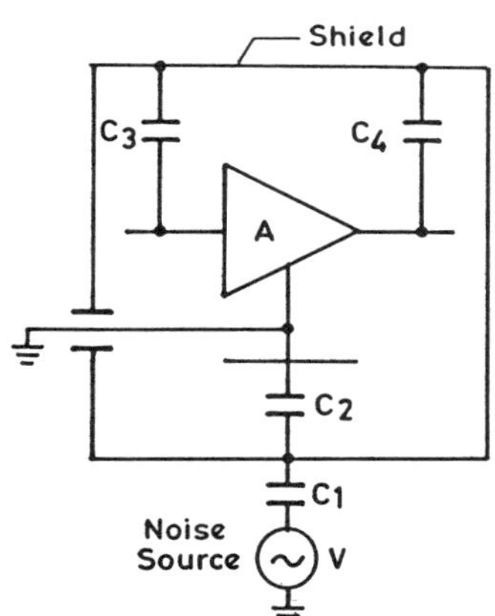

Figure 5.10a—Floating Shield with Coupling of Noise

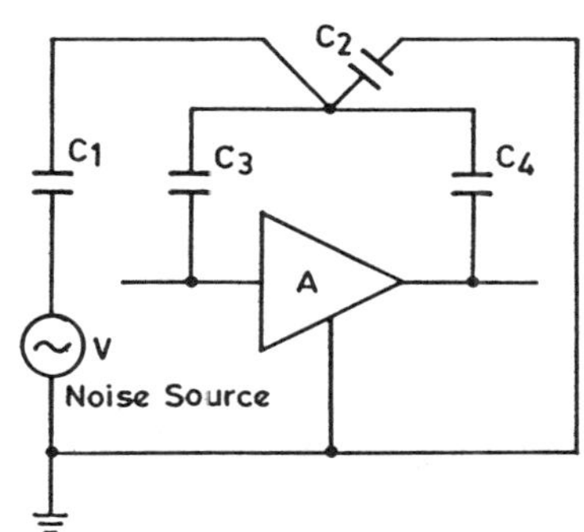

Figure 5.10b—Equivalent Circuit for Fig. 5.10a

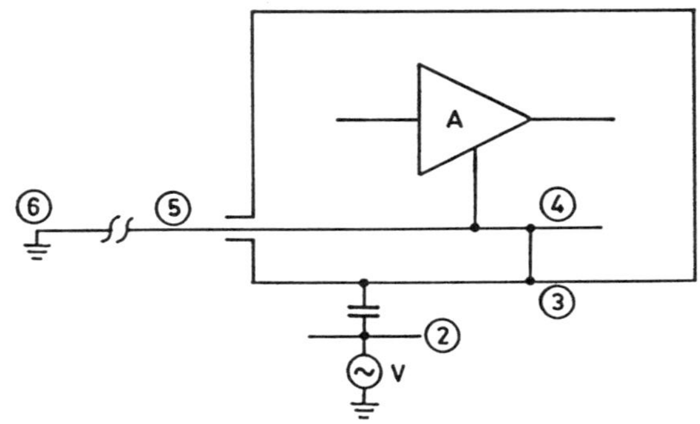

Figure 5.11—Grounding with Multiple Connections

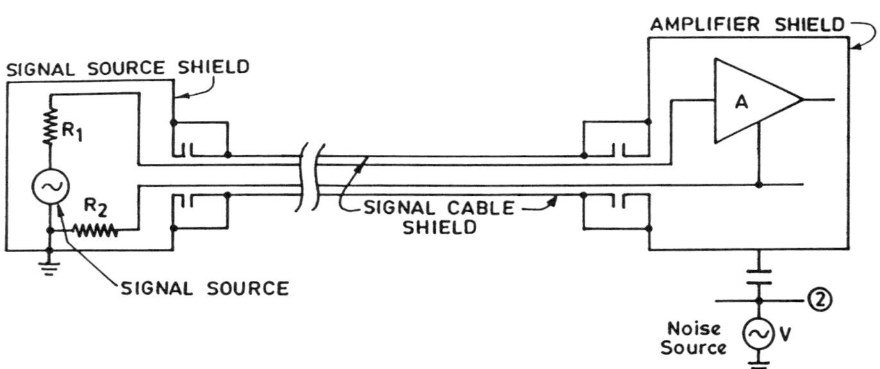

Figure 5.12—Grounding with Single Connection

5.8.3 The Problem of Shielding in Systems with More than One Ground

The rules given above require grounding of shields to the signal ground. However, a signal associated with one ground is sensed and a conditioned version of this signal is terminated at a second ground. These separate grounds can have potential difference between them. Shielding and grounding in this situation must be done in such a way that this potential difference does not contaminate the signal. Figure 5.13 shows how this can be done. The amplifier in this figure is a differential type. It is seen that the voltage between the two grounds appears as a common mode signal for the amplifier and is rejected. In order to avoid contamination, shield (3) must enter the amplifier to protect signal leads (5) and (6) so that the capacitance between each of these leads and the amplifier shield (4) is held below 1 or 2 pF. The amplifier input connector should be such that this is achieved. Clearly, a differential amplifier is essential here, and a high performance instrumentation amplifier should be used in this situation. These amplifiers reject common mode signals and amplify only the differential mode signals. The common mode rejection ratio, CMRR, is a figure of merit of these amplifiers. Good instrumentation amplifiers have CMRR of the order of 100 dB to 120 dB at 50 or 60 Hz and unity gain. This performance usually degrades by 20 dB per decade as a function of frequency and gain, e.g., a CMRR of 120 dB at 100 Hz will degrade to 100 dB at 1 kHz for the same gain.

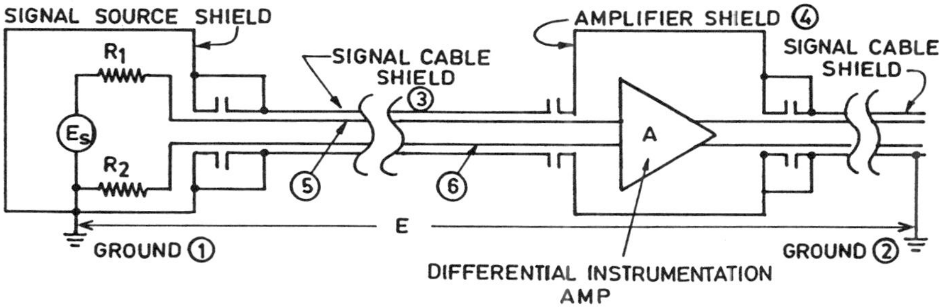

Figure 5.13—Grounding with Multiple Connections with Minimum Noise

5.8.4 Isolation Transformers

Transformers are built into most signal conditioning equipment. Shields around their primary and secondary windings and proper grounding of these shields can effectively eliminate the common mode noise signals that can enter through power lines. Thus, these transformers serve the dual purpose of noise isolation and power supply to signal conditioning equipment at appropriate voltage levels. Grounding and shielding of isolation transformers are discussed in References 4 and 5.

5.9 Recent Advances in Sensors [11–18]

Several technological developments have led to major advances in sensor technology. These have been primarily in the areas of semiconductor devices, microelectronics, and fiber optics. Some recent developments in sensors are discussed below.

5.9.1 Smart and Dumb Sensors [11–14]

Smart sensors employ a computing element, such as a microprocessor, in them. This element and the associated circuitry are housed in the sensor enclosure. Smart sensors thus employ computer techniques for direct and inferential measurement, computation, signal processing, and encoding. Smart sensors can be represented as shown in Fig. 5.14. There is much developmental activity in smart sensors and many are already on the market, especially for temperature, pressure, and flow measurement. It must also be mentioned that there is similar activity in the development of smart actuators. Sensors and actuators that are not smart will be referred to as dumb.

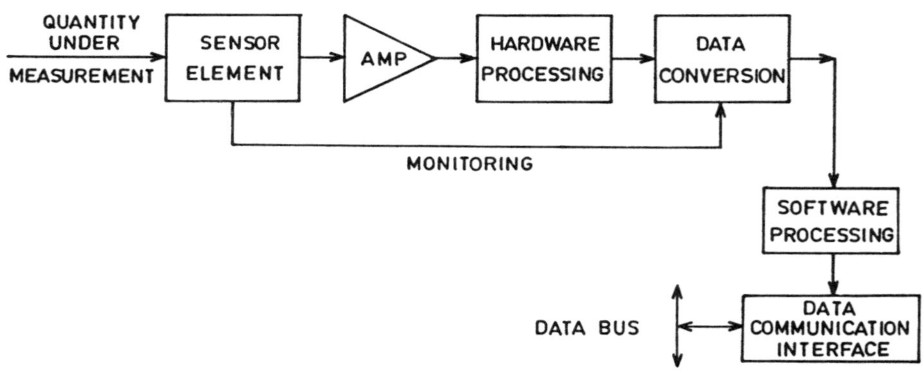

Figure 5.14—Smart Sensor

Advantages are gained in terms of improvements in signal-to-noise ratio and essential sensor characteristics (such as linearity, drift and cross-sensitivity) by doing signal conditioning and digital processing right in the sensor chip or housing. Furthermore, features such as self-test, drift cancellation, automatic calibration and range change, and addressability from a remote location are easily achieved. Multiplexing signals from various sensors is easy since the signals are digital. A major cost element in large and complex instrumentation systems used to be cabling for signals. This cost has come down significantly since signals can now be multiplexed and individual sensors addressed at will. Digital signals are immune to errors from channel nonlinearity and largely immune from induced noise, particularly if redundant coding is employed.

From the foregoing, it is clear that, due to the developments taking place in smart sensors, it is necessary to look at sensors in a new way, as subsystems capable of quite complex behavior. Smart sensors and actuators in process systems represent distributed measurement and control, with intelligence at each level.

5.9.2 Examples of Some Recent Developments in Sensors [11, 13, 15–18]

Monolithic Silicon Pressure Sensors

Crystalline silicon has some unique mechanical properties: it is harder than most metals; though brittle, it has a higher elastic limit than steel; it offers near perfect elasticity, which remains unchanged under repeated cycles of tension and compression. These properties are highly suited for making sensors for pressure, acceleration, and force, among others. Employing integrated circuit techniques and the recent advances in micromachining, new, inexpensive, high-performance sensors, which are smart to varying degrees, have been developed for batch production. Micromachining uses chemical etching techniques to form a variety of shapes in silicon wafers. From these, microsensors for force, pressure, acceleration, etc., can be made.

The most commonly used silicon microsensors are for pressure, of which there are basically two types: piezoresistive and capacitive. Both use a diaphragm 10 to 50 micrometers thick etched in a silicon wafer.

In the piezoresistive-type pressure sensor, four small areas at the opposite edges of the diaphragm are doped to form resistors. The four resistors thus formed are connected together to form a Wheatstone bridge. When the diaphragm flexes, two resistances increase and the other two decrease. Bridge output is converted into an FM signal by the circuit on the chip.

Different types of integrated circuit pressure sensors in different ranges up to about 400 bar are available. In the fully compensated sensors, on-chip calibration of offset and span and on-chip temperature compensation for temperature variation in the range –40°C to 125°C are provided. Factory laser trimming makes replacement chips and original chips interchangeable without the user performing any additional trimming or recalibration.

The capacitive type of pressure sensor is more sensitive than the piezoresistive type. Further, it is less sensitive to temperature changes. The diaphragm forms one plate of the capacitor. The flexing of the diaphragm due to pressure change induces a change in capacitance. This, in turn, causes a shift in the frequency of an associated oscillator. The capacitance change is very small. In order to avoid this change being masked by cable and stray capacitances, it is essential to have on-chip frequency signal generation and amplification. Many manufacturers produce these sensors for a large number of pressure ranges.

Fiber Optic Sensors for Temperature

Optical fiber thermometry is based on Planck's theory of blackbody radiation. A thin film of iridium is sputtered onto the end of a thin sapphire crystal to form a blackbody cavity. The radiant energy emitted by the cavity is characteristic of the temperature to which it is exposed. It can be transmitted almost loss-free through a fiber optic cable to a photodetection device. Planck's equation is used to determine the temperature of the blackbody.

Optical fiber thermometry is used in the range 500 to 2000°C. Accuracy achievable is 0.0025%. This represents a hundredfold improvement over the best accuracy with thermocouples. The thermal inertia of the blackbody is very small, enabling measurement of temperature changes with a bandwidth of up to 50 kHz. It is insensitive to electromagnetic interference and there is no thermal aging. Commercial units, complete with temperature sensor, optical detector, and a signal processor that can be interfaced to a computer through the IEEE 488 or RS-232C buses, are available.

Relative Humidity Measurement

Integrated circuits along with a polymer thin-film capacitor whose dielectric constant varies linearly with humidity are available for humidity measurement in the range 0 to 100 percent relative humidity. These operate in the temperature range –40 to +115°C and offer about 3 percent accuracy.

Flow Measurement

Differential pressure/orifice flow meters have been for many years the mainstay for flow measurement. Though they are not liked because they form an obstruction in the fluid flow path and are subject to wear if used with fluids that have suspended particles, they are simple, do not occupy much space, and are easy to maintain. Orifice plates can easily be changed. Recent developments in smart differential pressure sensors have given a new lease on life to this method of flow measurement.

Several companies manufacture smart magnetic flow meters. Improvements in coil design have resulted in lower power consumption and smaller size. Flow of fluids with conductivity as low as 1 micromho/cm can be measured. Magnetic flow meters with built-in signal conditioning and a microprocessor for calibration, automatic gain adjustment, and instantaneous digital indication of flow are now available.

Ultrasonic flow measurement technology is undergoing rapid refinement. Invasive meters are preferred at the present time. Though invasive, they provide very little obstruction to flow. Some ultrasonic flow meters use impeller wheels whose speed is sensed by ultrasonic means.

What has been discussed above is, by necessity, of an introductory nature because of the vastness of the subject matter. For details, the references cited here and other specialized texts should be consulted.

References

[1] Olive, Theodore R., and Danatos, Steven. June 1957. "Chemical Engineering's Guide to Process Instrument Elements." *Chemical Engineering*. pp. 287–320.

[2] Fribance, Austin E. 1962. *Industrial Instrumentation Fundamentals*. New York: McGraw Hill.

[3] Cerni, R. H., and Foster, L. E. 1962. *Instrumentation for Engineering Measurement*. New York: John Wiley.

[4] Bouwens, A. J. 1984. *Digital Instrumentation*. New York: McGraw Hill.

[5] Morrison, Ralph. 1984. *Instrumentation Fundamentals and Applications*. New York: John Wiley.

[6] Doebelin, E. O. 1983. *Measurement Systems (International Students Edition)*. New York: McGraw Hill.

[7] Henry, R. W. 1978. *Electronic Systems and Instrumentation*. New York: John Wiley.

[8] Boros, A. 1985. *Electrical Measurements in Engineering*. New York: Elsevier.

[9] Morong, B. July 3, 1980. "Isolation Stretches the Bandwidth of Two-Transformer Designs." *Electronics*.

[10] July 3, 1980. "Isolation and Instrumentation Amplifiers: Designer's Guide." *Electronics*.

[11] Browne, M. A. 1985. "Report on the Conference on Solid State and Smart Sensors held at IEEE Headquarters, London, 14 May, 1984." *J. Phy. E: Sci Instrum*. V. 18: 883–884.

[12] Brignell, J. E. 1985. "Interfacing Solid State Sensors with Digital Systems." *J. Phy. E: Sci. Instrum.* V. 18: 559–565.

[13] Jordan, G. R. 1985. "Sensor Technologies of Future." *J. Phy. E: Sci. Instrum.* V. 18: 729–735.

[14] "Sensors within Systems." 1985. *J. Phy. E: Sci. Instrum.* V. 18: 757–765.

[15] Babb, Michael. Jan. 1986. "Chips to Transmitters: New Pressure Technology Gets Going." *Control Engineering*. V. 33: 64–66.

[16] Blickley, George J. Nov. 1984. "Developments in Sensors and Measurement." *Control Engineering*. V. 31: 66–68.

[17] Blickley, George J. August 1985. "Flow Technologies: Not Standing Still." *Control Engineering*. V. 32: 77–79.

[18] Babb, Michael. Nov. 1985. "Micromechanics, Fibre Optics: New Sensor Performance." *Control Engineering*. V. 32: 54–56.

6
Buses and Communication

Sanjay K. Bose

6.1 Introduction

A typical computer system is required to communicate (i.e., send and receive) data with other computer systems or with various other instruments, sensors, transducers, actuators, or other devices that allow it to get data inputs from the external world for its processing requirements or send outputs that are generated as a result of such processing. These interfacing techniques are discussed in this chapter.

In addition, the ability of computers to communicate with each other for information exchange has become necessary. This caters to various requirements, starting with mundane tasks such as rapid transmission of messages and memos and continuing to the need for sharing resources (both software and hardware) and data across machines.

Such interfaces can be classified as follows:

1. Low-cost interfaces for transferring data at relatively low speed

2. High-cost interfaces for relatively high-speed data transfer

The first type of data transfer operation usually works slowly, typically one byte at a time. Here, the sending device or system will typically write one byte of data for the other device (or system) to read. Only after this byte has been read will the sender send the next byte of data to that interface. This continues until all the required bytes have been transferred. Under normal circumstances, input/output ports will be associated with each such interface, with the output port on the sending side and the input port on the receiving side; in situations in which either the sender or the receiver is not a computer system, the corresponding port may not be required. Moreover, such interfaces usually do not transfer a large volume of data; for example, in order to set or read a remote-controlled instrument, only a few bytes need to be transferred in either direction. On the other hand, a typical controller in a laboratory environment may have to take care of a large number of such interfaces in order to fulfill its input and output requirements. These interface requirements are fairly well met by the parallel and serial interfaces described in 6.2. Even though it is stated that interfaces of this type need not be very fast, it goes without saying that their speed should be as high as possible, since it may determine the overall speed of the system operation. Even if this is satisfactory, the system may be required to interface with a high-speed device for reading or writing data, which may impose speed requirements of its own.

The second type of interface requirement usually calls for more sophisticated hardware and software to allow large volumes of data to be transferred quickly in an error-free fashion between different computer systems that may not be in close proximity to each other. This is the function one expects from a reasonably good computer network (i.e., a LAN or local area network). A further advantage of such networks is that they provide multipoint connectivity between the systems they interconnect. This gives the user much greater freedom in deciding the source and the destination of the data. The needs of intermachine communication can hardly ever be met by the other kind of interface. Section 6.3 introduces the basic concepts of networking, and 6.4 discusses Ethernet-based LANs in detail.

Figure 6.1 shows the schematic of a typical system and the type of interfaces it may have. A network connection that will allow the system to communicate with other computer systems through a high-speed data network (e.g., ETHERNET or MAP) is shown. In addition, a number of parallel and serial interfaces are shown; some are intended to connect to other computer systems, while others are meant for interfacing sensors, transducers, actuators, or other peripheral equipment to this system itself. Some of the interfaces shown are serial and the others, parallel. As per the system's requirements, some of them may provide inputs to the system, while the others may be used for output from the system.

6.2 Interface Types

For the first type of interface mentioned earlier (low data transfer rate, relatively low speed), two are commonly used: the parallel interface and the serial interface. The choice between the two may, on occasion, be decided by the nature of the other equipment connected to the computer system; a common example is the printer interface, which may be either serial or parallel in nature depending on whether a serial or a parallel printer is to be connected to the system. In certain situations, the choice may be dictated by the interfaces that are available with the computer system. Most computers will have a serial interface that is capable of bidirectional data transfer but may not have a similar parallel interface; an example is the PC which typically provides two serial ports to the user but only one (or sometimes two) parallel port, which may be used only to output data (typically to a parallel printer or similar device). In systems and applications in which the user does have the complete freedom to choose between a serial or a parallel interface for input or output, the choice is usually dependent upon the speed at which the data transfer is required and the

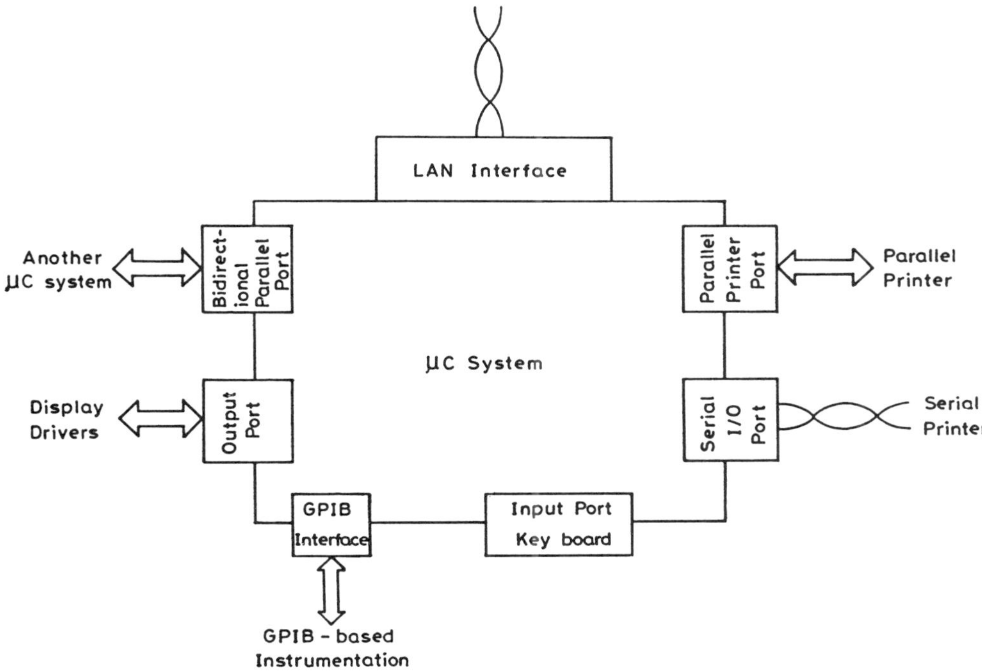

Figure 6.1—Interfaces in a Computer System

physical proximity of the computer system and the equipment to be interfaced. If the distance between the two is large, a serial interface is generally preferred because it requires fewer lines to be connected between the two; this, however, should be carefully balanced against the fact that a serial interface is slower at transferring data than a parallel interface and, in noisy environments, may even be subject to a greater number of errors. In the following two sections, these two types of interfaces are described in greater detail.

6.2.1 The Parallel Interface

The parallel interface essentially provides a way to transfer several bits of data (typically a byte or eight bits), in a parallel form, to or from the computer system of interest. Unlike the serial interface, where multibit data is transferred one bit at a time, the parallel interface transfers all the bits (in a byte) simultaneously, thereby allowing greater speeds of operation. Unfortunately, this would require several parallel lines to actually carry the data. This tends to make such an interface costlier (and somewhat more clumsy) than a serial interface, especially when the distances involved are reasonably large. To make matters worse, one would also need additional lines for "handshaking", i.e., establishing a proper data transfer protocol between the source and the destination so that the data is not lost or overwritten; this would further increase the number of lines required by a parallel interface.

To the computer system, the parallel interface appears as a digital port, which may be either an input port or an output port. Input ports are used to transfer data from systems or devices outside of the computer system of interest; output ports are used to transfer data in the reverse direction. Such a port will typically be able to carry eight bits (or one byte) in parallel between the data bus of the computer system and the external device. One can also consider having bidirectional I/O ports that can transfer data simultaneously in both directions. Such ports are somewhat more difficult to interface (software and hardware) than straightforward input and output ports but are ideal for applications in which the interface is expected to handle bidirectional data transfers between two computers or microprocessor-based systems.

It was mentioned earlier that parallel ports may require other control signals for handshaking in order to effectively set up a data transfer between the source and the destination of the data. As shown in Fig. 6.2, for any port, input or output, these signals may be categorized as follows:

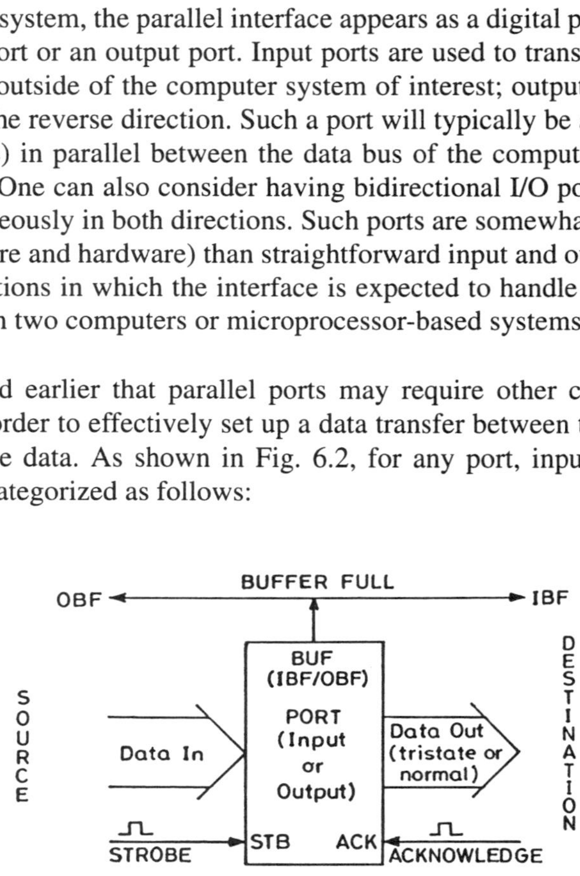

Figure 6.2—Data and Handshake Lines for a Port

1. *Strobe.* A pulse at this input is used to latch data into the port. For an output port, the computer system should provide this pulse along with the data. For an input port, this would be provided by whatever system or device wants to transfer data to the computer (once again, along with the data). In device literature, this pulse may be referred to as STB and may be either active high or active low, depending on the device.

2. *Buffer Full.* This is an output from the port indicating that the port (or buffer) is currently full, i.e., data last written into this port has not yet been read by the

destination. For input ports, this is commonly referred to as IBF (input buffer full); for output ports, this may be referred to as OBF (output buffer full). Apart from indicating to the source device that the last data written has not yet been read (hence, further writes may cause data to be overwritten and lost), this may also be used to indicate to the destination device that this particular interface has data ready to be read. In interrupt-driven applications, this signal may be appropriately interpreted as an interrupt by either the source or the destination.

3. *Acknowledge.* This is an input to the port from which the data is read. A pulse at this input should be issued whenever the destination device wants to read the data latched on the port by the previous strobe signal. For applications in which the output of the port is tri-state (i.e., for connecting to a bus), the output will appear on the otherwise floating output lines of the port only when this pulse is present. For situations in which the output may be continuously ON (tri-state outputs are not needed), this pulse is still needed to signal to the port that the previously written data has been seen by the destination device. This signal effectively clears the buffer full condition set by the strobe, indicating that the buffer is now empty. In typical device literature, this signal may be referred to as ACK.

It should be noted that for a particular device or port, the names of these signals and their polarities—active high or low—may be different. It is also possible that the work of a particular signal is effectively done by a combination of two or more signals. A separate interrupt (or INT) may also be provided with slightly different timing compared to the buffer full signal; however, these two signals perform essentially very similar functions.

Figure 6.3 gives the timing diagram of the data lines and the handshake signals as they are normally used. Note that the strobe signal is applied with valid data inputs and latches this data in the port. It also sets the buffer full signal, which should be supplied as the IBF and OBF inputs to the destination and source systems, respectively. This signal is reset once the ACK signal is applied by the destination device to read this port. At this point, the source can write the next byte of data to this port without fear of overwriting. Note that for certain ports (typically those with tri-state outputs), the data output from the port (to be read by the destination) may be valid only during the time the ACK signal is applied; for such ports, the destination should not accept the output of the port to be valid data at other instants.

Consider the port shown in Fig. 6.2 once again. If this is to be used as an input port for the computer system, the system itself should occupy the position indicated by DESTINATION and the external device would be the SOURCE. This direction will be reversed if this port is to be used for output from the system.

It has been mentioned earlier that a bidirectional port may be used as an interface if data is to be transferred in both directions. The easiest way to visualize such a port is

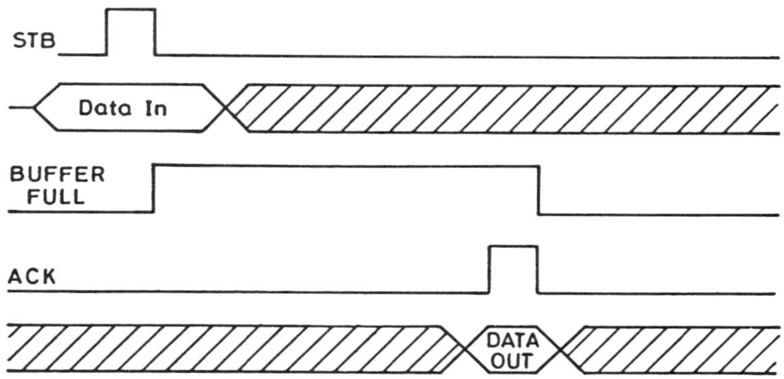

Figure 6.3—Data and Handshake Timing for a Port

to consider it to be formed using two ports as shown in Fig. 6.2, where the two of them are connected in parallel but in opposite orientations. The handshake signals for such a port will then consist of a pair of the set of handshake signals, shown in Fig. 6.2, with one set for each direction of data transfer.

6.2.2 The Serial Interface

The serial interface provides an alternative method for connecting two computer systems or a computer system with an external device. As mentioned earlier, its biggest advantage compared to a parallel interface is lower cost, because fewer lines are needed for this connection; however, it cannot provide speed of operation as high as a parallel interface. Serial interfaces tend to be cost effective for distances of 100 meters or more.

Serial data transfer essentially consists of parallel-to-serial conversion at the transmitting end and a serial-to-parallel conversion at the receiver end following the reception of the serial bit stream. The data bits are sent at a fixed rate that is called the bit rate or baud rate. Special bit patterns may be needed to signal the start or end of a data word (typically an 8-bit or 7-bit data or character) to the receiver. Based on this, two modes of serial transfer are usually recommended. In the *asynchronous* mode, the special bits are required to signal the start and end of each word; this wastes a lot of overhead and also reduces the overall rate at which transfer can be done. This, however, is the most common and standardized method of establishing a serial connection and is also the cheapest and easiest to implement. We will describe this and its associated standard, the RS-232C interface, in greater detail subsequently. For high-speed serial data transmission, the *synchronous* mode of transfer is usually recommended. This does not incur the overhead mentioned above for signalling start and end (of word) but imposes the additional requirement that if a data block is to be sent, it must be sent continuously until the end without any idling in between. (This requirement is not there for the asynchronous mode, where the transmitter can idle between data words.)

In order to guarantee the integrity of the data received, these serial transfer methods also incorporate some type of error detection. For the synchronous methods, this is usually done by adding a 16-bit or 32-bit cyclic redundancy check pattern to each data block following CRC coding (this is usually built into the communication interface devices used). The asynchronous method can also use such CRC coding; the accepted practice, however, is to append a parity check bit to each data word (7-bit or 8-bit) being transmitted; this is then checked by the receiver to ensure that the received data is correct. This asynchronous serial transfer technique is discussed in detail next.

The format followed for transferring a data word using the asynchronous serial technique is shown in Fig. 6.4. Note that, in this figure, the two signal levels are mentioned as being MARK and SPACE. In practice, one will use a high level for one of them and a low level for the other; needless to say, for communication to be effective, the sender and receiver must make this choice in a mutually consistent fashion. The idle state of the line is a marking condition. The transmission of a data word is started with the START bit, which is a SPACE; this MARK-to-SPACE transition is also taken by the receiver as the signal indicating possible start of data (most receivers would cross-check this by re-examining the START bit in the middle to confirm that it is indeed a SPACE). This is then followed by n bits of data corresponding to the n-bit word that has to be transferred serially; typical choices for n are 7 or 8, but $n = 5$ or 6 is also possible (of course, the receiver and transmitter must make consistent choices).

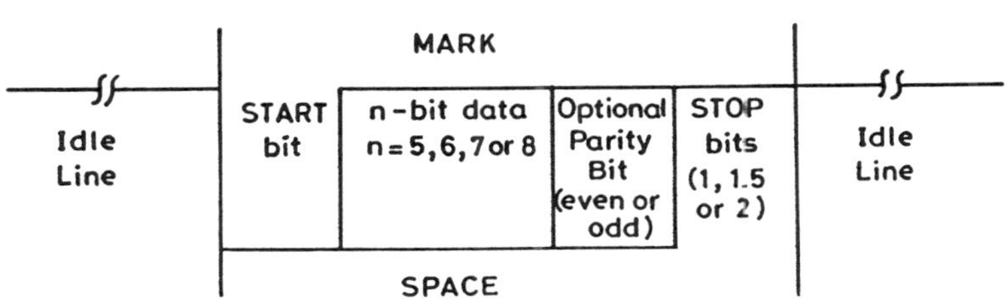

Figure 6.4—Asynchronous Serial Timing Format

After these data bits have been sent, the transmitter may send a parity bit for error detection purposes. This bit is optional, and, if parity is selected by both the sender and the receiver, one can further choose the option of using even or odd parity (i.e., if even parity is selected, the parity bit is such that there are an even number of 1s in the n-bit data and the parity bit). The data transmission for this n-bit data word is then terminated with STOP bit(s); the transmitter has the option of choosing one, one-and-a half, or two stop bits, but the receiver usually checks for one STOP bit. After the STOP bit has been sent, the transmitter can go idle or may decide to send the next data word in the same format once again. It should be noted that the conventions used for the idle state, START bit, and STOP bit(s) are basically required to signal to the receiver when the transmission of a data word begins and when it ends. It is, of course, expected that the receiver and transmitter are using a compatible set of choices (e.g., choice of n or parity) and are using the same bit clock for clocking the serial bits in and out. Note that, as the name implies, the asynchronous serial interface does not carry the actual bit rate clock from the transmitter to the receiver; it is not difficult to see that as long as their local clocks are reasonably close in frequency, this would also not be needed.

Various commercially available LSI devices may be used to provide a serial I/O interface to a computer system. These devices usually take care of the parallel-to-serial conversion and data transmission at the sending end and the data reception and serial-to-parallel conversion at the receiving end. The computer system communicates with this device by treating it as a special kind of port that actually provides four ports (at least!) to the computer system. One of them is an input port from which the serial data received is read, and another is an output port into which data to be serially transmitted is written. In addition, the device may have two more ports: an input port for the computer to read back status information from the device and an output port that may be used to send special commands to the device or to control handshake signal outputs discussed later. Figure 6.5 shows a typical interface for such a device.

In the figure, the serial I/O device interfaces to the computer system on one side and to the serial I/O device of the other system or the other device. The two serial I/O devices may either be connected directly or through two appropriate modems, one on each side. On the serial connection side, the TxD and RxD lines are used, respectively, for data transmission and reception. The other lines are used to handshake data transfers between the two serial devices. The signal CTS (Clear to Send) is an input to the device, indicating that it can send data (if any) to the other device. This will normally be driven by the RTS (Request to Send) output of the other device; the serial device outputs a high on RTS when it is able to receive data from the other side. The DTR (Data Terminal Ready) and DSR (Data Set Ready) lines are similar except that they are used when the serial device uses a modem for communication. By convention, the serial device outputs DTR high to signal the modem that it is

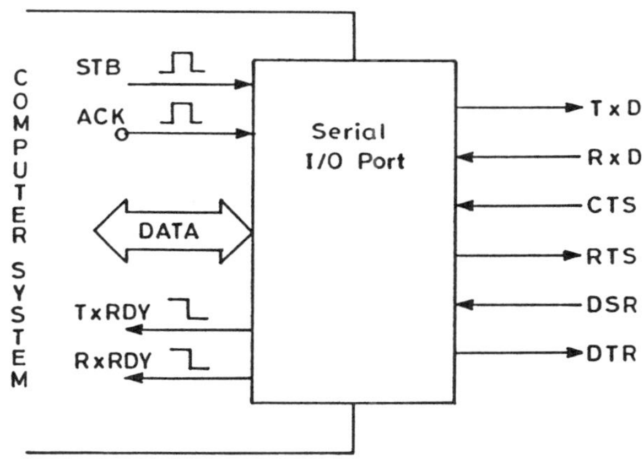

Figure 6.5—A serial Input/Output Interface

ready; the modem, in turn, applies a high to the DSR input of the serial device when it is also ready.

For the interface on the computer side, the serial device connects to the data bus of the computer system in the usual fashion. The computer system uses the STB signal to latch data into the device; these could either be commands or data to be serially transmitted. It also uses the ACK signal to read data (received data or status) from the device. (Note that the actual data is, of course, passed through the data bus connection.) In addition, the device may have two special outputs, TxRDY and RxRDY, that may be used appropriately by the computer system either as buffer full/empty indicators or as interrupt inputs to the computer. The TxRDY input signals the computer that the serial device can now accept a fresh data word (typically a byte) from the computer for serial transmission (i.e., without fear of overwriting the previous byte written for transmission); in a way this is the buffer empty signal for transmission purposes. The RxRDY output, on the other hand, is a buffer full indicator, signalling that the last data byte received from the serial line RxD has not yet been read by the computer; the computer has the responsibility to respond as fast as possible to the RxRDY=1 condition and read the receive buffer before the data there gets overwritten by subsequent data received over the serial line. This line is very often used as an interrupt signal that triggers the computer to read the received data.

Two interconnected devices cannot both transmit on the same line. The output of one must be connected to the input to the other. This is done by dividing devices into two categories: data terminal equipment (DTE) and data communication equipment (DCE). RS-232C connections are brought out on a 25-pin D-connector. Most AT class machines use a 9-pin subminiature D-connector in place of at least one of the

25-pin connectors. DTEs have male connectors and DCEs, female connectors. However, some suppliers do not always follow this convention. The ports on a PC are usually configured as DTEs. The connections of RS-232C (DTE) are given in Table 6.1. On the IBM PC and some clones, +5 V is available on pin 10. This is a nonstandard feature. On a DCE, the pinouts of the TxD-RxD, RTS-CTS, and DTR-DSR pairs are reversed. Two DTEs or DCEs can be connected together by using a special cross-wired cable (null modem), as shown in Table 6.2. It is obvious that, in a null modem, the three pairs of pins mentioned above are cross-wired and the grounds connected straight through. Such cables can be easily made up or purchased. On the PC using RS-232C, baud rates up to 2400 can be safely used under most conditions. For higher rates (up to 9600 baud), it may become necessary to write special drivers. A modification of the RS-232C known as the RS-422 is also becoming popular. The wiring is similar but the signals are differential. RS-422 can support higher baud rates and longer cable lengths. Also, many devices can be connected in parallel on RS-422, which is making it popular as a local instrumentation bus for connecting controllers to the host computer. Even though RS-232C has been updated to RS-232D in 1988, the term RS-232C is commonly used to refer to the entire RS-232( ) family.

Table 6-1—RS-232C Pin Connections

Pin	Signal on DTE	Signal on DCE
1	Chassis Ground	Chassis Ground
2	[3] Transmitted Data (TxD)	Received Data (RxD)
3	[2] Received Data (RxD)	Transmitted Data (TxD)
4	[7] Request to Send (RTS)	Clear to Send (CTS)
5	[8] Clear to Send (CTS)	Request to Send (RTS)
6	[6] Data Set Ready (DSR)	Data Terminal Ready (DTR)
7	[5] Signal Ground	Signal Ground
8	[1] Data Carrier Detect (DCD)	
20	[4] Data Terminal Ready (DTR)	Data Set Ready (DSR)

Note: DCD is used with modems only. Pins on 9-pin D-connectors are shown in parenthesis in DTE signals

Table 6-2—Null Modem Cable Pin Connections

Connector 1	joins to	Connector 2
1	—	1
7	—	7
2	—	3
3	—	2
4	—	5
5	—	4
6	—	20
20	—	6

In RS-232C there are many variations, making this interface notorious for problems in making it function. While on the Centronics® there is a better than 90% chance that the interface will work the first time round, on the RS-232C it may sometimes take many attempts to make it function properly. The MODE command of DOS is used to set the baud rate, parity, 7- or 8-bit word length, number of stop bits, etc. Even with everything apparently correct, sometimes the interface still does not work. In such a case, the best method is to disable the handshakes and then proceed stage by stage until the interface works properly. To disable the handshakes, the easiest way is to short the pairs of pins 4–5, and 6–20 at either end. The interface should be run at the lowest baud rate available. Once it is functioning properly (it may lose some data due to the disabling of the protocol but this is normal), the handshakes are connected one at a time. Once the interface is working properly with the handshakes (there should be no lost data now), then the speed can be increased.

6.3 An Introduction to Computer Networks

Recent advances in semiconductors have provided designers with sophisticated processing capability at very low cost. Based on LSIs and VLSIs, these systems are providing users with an unprecedented level of computing power and versatility. Instead of investing in a single costly minicomputer or mainframe, the current trend is to dedicate low-cost microcomputers for each application. Apart from cost factors, other considerations, such as reliability, maintaining backups, or fast information dissemination, may also recommend such a distributed approach.

A distributed system with processors at different locations or *nodes* would typically need the services of a network to connect them together and allow data transactions between them. With cost-to-computing ratios as they are at present, a mainframe may cost 100 to 1000 times more than a microcomputer but may only be 10 to 100 times faster. It is, therefore, cost-effective to set up a system with separate micros for different applications with a network interconnecting them. Apart from allowing centralized monitoring and control, such a network will then also permit costly peripherals to be shared over the network. This resource sharing should allow for more effective utilization of these peripherals. Reliability is also a major factor motivating networking. In contrast to a system with one major mainframe, failure of a few nodes or micros in a network may still allow the remaining nodes to function. Finally, a computer network may also serve as a medium for the rapid dissemination of information. In this context, it can provide services such as electronic mail and lead to a greater degree of office automation.

In order to standardize the design of networks and encourage a modular approach, the International Standards Organization (ISO) has standardized a seven-layered architecture called the Open System Interface (OSI). This ISO-OSI model divides a network into seven layers, in which each layer performs a well-defined function. Viewed across the network, each layer presents a different level of abstraction, which is transparent to the lower layers. This layered architecture is shown in Fig. 6.6.

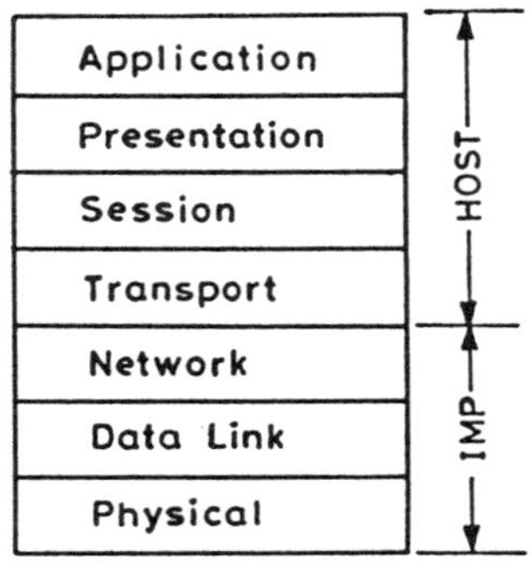

Figure 6.6—Layered ISO/OSI Interface Structure

The ISO-OSI model requires that the layers be present over the entire network. Typically, each node will have a host computer and an intermediate message processor (IMP). The IMP, also referred to as the network interface processor, actually interfaces the node to the network. Functionally, the hosts are required to talk to each other; physically, the host actually talks to its resident IMP, which, in turn, uses the network to talk to the other IMP. In a simple application, the same processor may act both as the host and the IMP; even in such cases the software functions of the two may still be clearly identifiable.

The various layers of the OSI model are as follows:

1. The *Physical Layer* is the lowest layer of the network and is concerned with the actual physical connection between the different nodes. Transmission of the raw data, i.e., 1s and 0s, modulation-demodulation, data format (NRZ, NRZI, Manchester encoding), and the mechanical and electrical interface to the network is the responsibility of this layer.

2. The *Data Link Layer* shields the higher layer from the characteristics of the physical transmission medium and provides an error-free and reliable data link connection. In other words, it passes raw binary data to the Physical Layer that it ultimately wants transmitted to the Data Link Layer on the other side. The data from the Data Link Layer is organized into frames or packets with additional error-detection and sequencing information. This is needed to guarantee error-free data transactions in the right sequence over the imperfect Physical Layer.

The Data Link Layer assembles the received frames in the right sequence before passing them on to the Network Layer.

3. The *Network Layer* provides for transparent transmission of data from a Transport Layer in one system to a Transport Layer in another system. This is divided into two sublayers: Routing and Virtual Circuit. In Routing a center is created for messages to be transmitted in a given form. The Virtual Circuit sets up the specified path from the source to the destination for the communication call; the right sequence is the one in which the source Data Link Layer received it from its Network Layer. These objectives are achieved by means of error detection schemes, sequence numbers for frames, and acknowledgments for correctly received frames following an appropriate ARQ protocol. The Network Layer subdivides the data it gets from its Transport Layer into packets for eventual transmission to the Transport Layer on the other side. Typically, the functions of the lower three layers are taken care of by the node's IMP; the higher four layers are the responsibility of the corresponding host.

4. The *Transport Layer* provides the functions necessary to bridge the gap between the Network Layer and those above. This layer is to complement the underlying network to ensure that the user get the required quality of service.

5. The *Session Layer* forms the user's interface to the network. It enables two applications to communicate across the network and performs the functions of security, name recognition, logging, administration, etc.

6. The *Presentation Layer* performs functions that are requested often enough by users so that a general solution may be provided by the network designer. In other words, it identifies and agrees to the common system to be used by two applications for communication.

7. The *Application Layer* is completely user-dependent and consists of the functions that the user wants.

Any LAN can be considered as being comprised of two parts, namely, physical and logical. The physical part basically concerns the transmission media, i.e., cable, taps, modems, interface cards, etc. The logical part is supported by the protocols and software routines.

The communication subnet of the network, shown in Fig. 6.7, essentially provides the connectivity between the various nodes of the network. This can be done either by providing direct connections between each IMP pair or through indirect connections via intermediate IMPs. The former is a very expensive solution unless the number of nodes is very small; the bus architecture is an exception to this. The latter approach is more viable. In most networks, the communication subnet typically consists of either point-to-point or broadcast links. In the former, the subnet contains direct links between the various IMPs with each link having its own unique source and destination IMP. Such networks tend to be partially connected, with intermediate IMPs needed to store and forward packets going from the source to the destination IMP. For networks with a broadcast link, the link is such that all the IMPs can transmit data on it or receive data from it. Such a network is fully connected and is also referred to as a bus type of network. Here, appropriate access mechanisms must be provided if more than one IMP wants to transmit on the link at the same time.

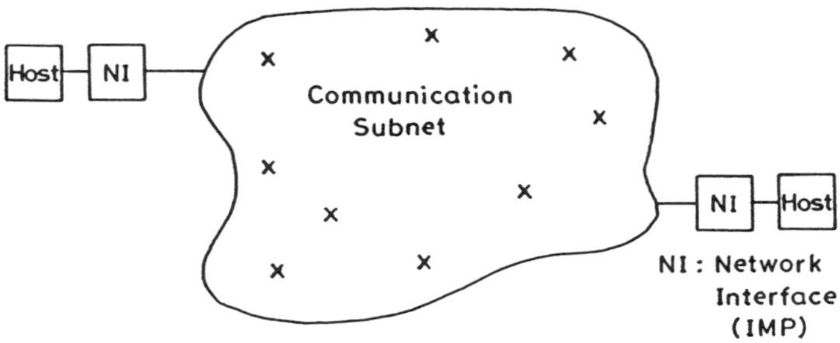

Figure 6.7—Schematic Representation of a Communication Subnet

The star requires a central node that is connected directly to all the other nodes. Communication between two nodes will have to be routed through the central node, which often acts as the network controller. For systems with a large number of nodes, the software and hardware of the central node becomes too complex to allow a low-cost network design. Moreover, this architecture makes it very difficult to add more nodes to the network if the need ever arises.

The tree offers another choice of subnet—one that has only point-to-point links. This topology is not popular except in applications in which a natural hierarchy exists among the various nodes.

For most LAN designs, the most popular choices are the ring and the bus topologies, since both of them allow users to be added and deleted in a very simple fashion. The ring network circulates data along the ring in one direction. The source introduces the data on the ring; this is circulated around the ring, repeated by intermediate IMPs, and finally reaches the destination. Transmissions on the ring may be controlled by a variety of techniques. In the token passing scheme, suggested by the IEEE 802 Committee, a special bit pattern (referred to as the token) is circulated around the ring from one node to the next downstream node. When a node gets the token, it does one of two things. If it does not have any data to send, it merely forwards the token to the next node. If it has data to send, it keeps the token, transmits all its data, and then forwards the token. Note that a node cannot send data unless it has the token and will then act merely as a repeater for the data it receives from the upstream node. Figure 6.8 shows the star, tree, ring, and bus topologies.

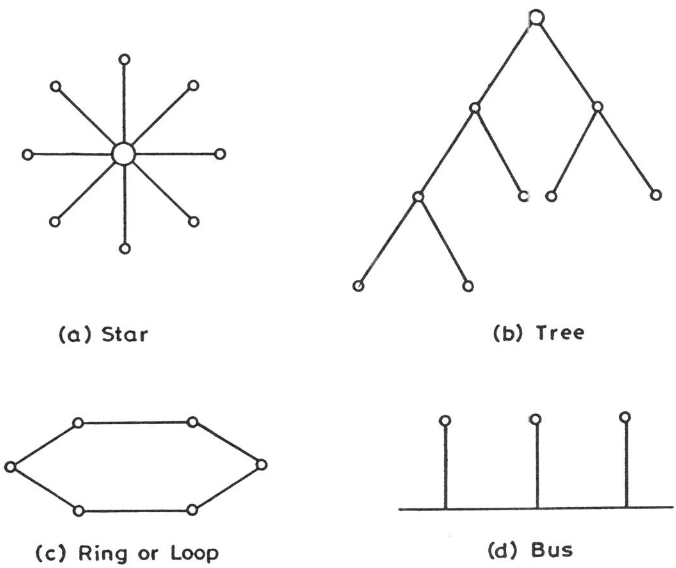

Figure 6.8—Some Network Topologies

A very popular bus architecture for LANs is the Ethernet and other similar systems. Here, the broadcast medium is a multidrop coaxial cable that is tapped by each node for transmission and reception. Access arbitration is done by the CSMA/CD protocol, which is also one of the recommendations of the IEEE 802 Group.

The ring or loop may cause greater delays in transporting data than the bus architecture because of the repeater action at each mode. This is especially true when the load is light. However, the CSMA/CD protocol of the bus architecture may show instability in contrast to the unconditionally stable behavior of the token ring. In addition, the network interface for the CSMA/CD is complex, is costly to implement, and requires good cabling. Because of these reasons, the token-passing ring architecture (ARCnet) was popular in low-cost network designs. However, Ethernet, which uses CSMA/CD, is the dominant networking system in office automation, with over 80% of the market.

The IEEE 802 Committee has proposed three standards for LAN architecture that are designed to fit the requirements of different types of applications. A brief description of each of the three standards is given below.

6.3.1 Ethernet, IEEE 802.3 CSMA/CD Protocol for LANs

This standard follows the Ethernet system developed by XEROX Corporation®. This LAN was the first to be standardized and is fairly extensively used in high performance computer networks today. It uses a bus architecture implemented through special coaxial cables. Different nodes interface to the network by connecting themselves to this cable through external "cable piercing" taps. A very large number of users may be connected at the standard data rate of 10 Mbps over a distance of up to 2 km.

The protocol used to access the network is carrier sense multiple access with collision detection (CSMA/CD). With CSMA/CD, a station that wishes to transmit listens before it actually begins transmission. Transmission starts only if the medium is actually sensed to be idle. It is possible that two stations start their transmissions together, which leads to a collision, causing data from each transmitting station to be corrupted. In that case, the transmitters themselves detect this collision, stop transmitting, reschedule their transmissions by a random delay, and try again.

A large number of standard computer systems now offer Ethernet for networking purposes. Several independent suppliers can provide cards that may be plugged into the computer to give an Ethernet network interface. This network is well supported, both in terms of its hardware and software as well as applications. It has been around for a long time, demonstrating its viability in supporting computer networking. A high-performance EISA Ethernet card is shown in Fig. 6.9.

6.3.2 MAP, IEEE 802.4 Token Bus Protocol for LANs

This protocol is fast emerging as the shop floor automation standard and, hence, is of great importance to people who are interested in networking complete factories. This protocol was initially developed by General Motors®. MAP is the acronym for manufacturing automation protocol. This also has a bus architecture like CSMA/CD but differs from it in certain basic features. Unlike CSMA/CD, here the right to access the token bus is controlled by means of a token that is passed around from one user to another. In this case, the token is a logical token, which is essentially a special message that gets passed over the bus from one station to another. The station that currently has the token is the one that has the right to transmit. Following its own transmission, it sends the token to the next station; if the station does not have anything to send, it merely forwards the token to the next station.

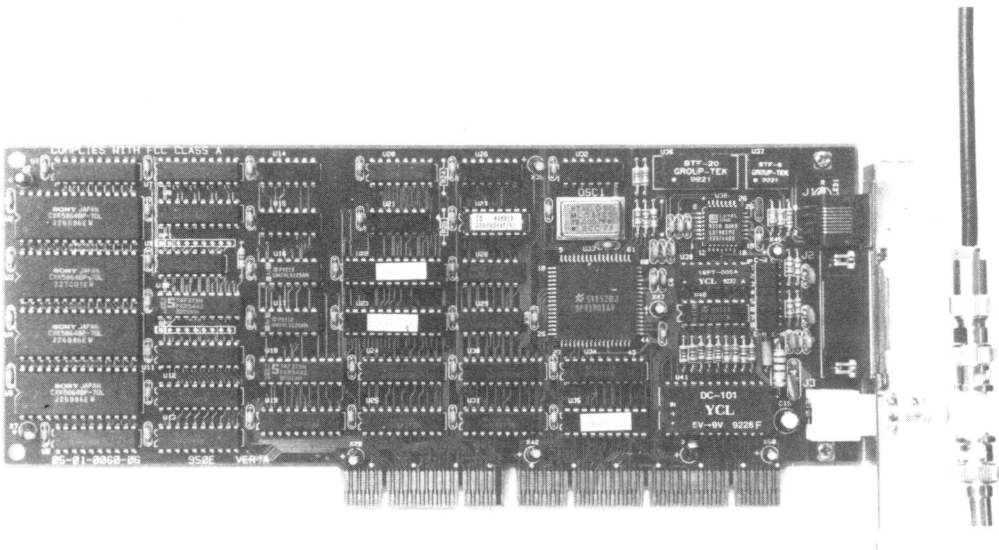

Figure 6.9—High-Performance EISA Ethernet Card with Cable, Tee, and Terminator

(Courtesy Zenith Computers Ltd.)

The transmission strategies of the token bus and CSMA/CD are also somewhat different: token bus uses a modulated carrier, and CSMA/CD directly sends high/low levels (i.e., at baseband). Actually, the full system specification for MAP allows for three different modulation schemes—two single-channel and one broadband. The single-channel modulation is for the token bus data; the broadband channel can carry a large number of signals multiplexed over the same cable. Single-channel implementations are simpler and cheaper; however, the broadband schemes permit longer distances, greater volumes of information to be carried, and greater versatility in exchange for significantly higher cost and complexity. It should be noted that the ISA PROWAY standard is very similar to the IEEE 802.4 recommendations with some minor modifications.

For shop floor automation, token passing bus is preferable to CSMA/CD for its constant latency. In a CSMA/CD system when the traffic is light, the bus is almost immediately available, i.e., the latency is very low. However, as the traffic increases, the availability of the bus slowly but progressively degrades, i.e., the latency increases. At very high traffic levels the bulk of the time is spent in collisions, with very little traffic actually getting through. In other words, in a lightly loaded network the latency is very low, and this increases with load; in a token ring system the token is guaranteed every so often. However, a token passing bus has a constant latency which is crucial for automation applications.

6.3.3 IEEE 802.5, Token Ring Protocol for LANs

IBM has been a major proponent of this system in recent years. Its operation is logically similar to that of the token bus except for the following differences:

1. It uses a ring architecture for the network (unlike the bus of CSMA/CD or token bus).

2. The various stations on the ring are actually connected by point-to-point links, where the transmit line of one station is connected to the receive line of the next downstream station. Extended to all nodes of the network, this essentially completes the ring.

3. Each node repeats all the data it receives to its next downstream node, thereby allowing the data to traverse the ring and be "audible" to all stations.

4. The transmitting station removes the messages it has inserted in the ring once they have made a complete transit of the entire ring, which is done to ensure that data inserted in the ring does not circulate there forever.

The token in token ring systems is actually a specific bit pattern rather than an actual message. The right to access is granted in the same way as for a token bus system. The token ring is also a baseband system, as is CSMA/CD. The other standards use a bus architecture, which requires the use of special and costly interfacing techniques.

In the area of office networking, ARCnet™ (which is a 2-Mbps token ring) has gained popularity as a less expensive and robust system. However, with the decrease in costs of Ethernet cards, the use of the ARCnet is declining rapidly, with the near universal adaptation of Ethernet, in both the bus (10 Base 2) and star (10 Base T) configurations.

6.4 Office Networking

The use of office networking systems has become widespread during the past five years. Their utility lies in interperson and interoffice communication as well in resource sharing. As software has become progressively larger and often more costly, the need for sharing this invaluable resource in terms of both cost and storage space has grown. This section deals with various aspects of office networks. It discusses peer-to-peer and client-server networks, the networked environment, and some popular PC LANs. Subsequently, the hardware of popular Ethernet configurations is discussed.

6.4.1 Peer-to-peer and Client-server networks

In peer-to-peer systems all systems are treated as equal. Various hardware resources are distributed across the network. For example, the network environment may utilize hard disks on various computers for software storage and retrieval; some of these nodes may also have CDROM drives, printers, etc.

In a client-server environment one or more of the computers on the network that are designated as the servers contain the resources (software as well as hardware) that are made available to the client machines. This system is logically of the hub and spoke type.

Both peer-to-peer and client-server systems have inherent advantages and disadvantages. In a client-server environment, typically there is a server, which is used to hold all the system software, along with printers, links to other systems, etc. The server is not used for any computing and is dedicated to serving the clients. Thus, the bulk of the cost in terms of large disks, backup devices, laser printers, and the like is centralized. Software maintenance is done at a centralized place, where periodic backups are also taken. By suitably restricting other nodes, the risks of virus infection can also be minimized. Furthermore, the server can be optimized for network performance. The client machines, on the other hand, can have minimal peripherals (such as disks and printers) to reduce cost. Using boot ROMs, diskless configurations are also available. The network software is similarly optimized and typically tends to be rather large on the servers with very small overheads on the clients. Internetwork connectivity is implemented through the servers.

A peer-to-peer system is more suitable for smaller networks. In this case, all nodes on the network are treated as (almost) equal. Software is distributed over the various nodes, and users on other nodes can access it. Obviously, the size of the server software has, perforce, to be limited to minimize overheads, since it has to be installed separately on all the nodes. Software maintenance is again distributed over various machines, and if the node carrying a particular package is not active, that package will not be available to the other users. On the other hand, the failure of any one node is not nearly as disastrous as the failure of the server in a client-server system.

It should be noted that the two are not mutually exclusive. Linkages between independent networks are possible and are indeed common. Quite often, peer-to-peer networks are operated inside the broader framework of a large client-server system. In such a system, all the machines are clients to the server(s); at the same time, some of them are mutually sharing resources in a peer-to-peer mode.

6.4.2 The Networked Environment

A networked system offers a far richer working environment than stand-alone machines. The facility of electronic mail allows messages to be sent electronically between users and also to other systems through gateways. Similarly, users can send and receive facsimile messages through the network. Many experienced users find this communication facility alone as more than enough justification for networking.

Any software will not be required by all its users at the same time. Through networks, it can be made available to the larger audience without having to have as many copies, thus saving on cost. For example, ten persons in an office may require access to a particular word processor at some time or another. However, it is very unlikely that all ten will wish to access it at the same time. Thus, a network license for a fewer number of users may suffice. As single-node software is generally not licensed for network use and software licensing conditions vary, users should check before placing any software on a network. Furthermore, modern software very often requires a vast amount of system resources. Packages that require well in excess of 10 MB for installation are commonplace, and some packages have grown to the extent that they are shipped on CDROMs. The resource requirements alone make networking almost mandatory in such cases.

Security has always been the bugbear of stand-alone computing. Personal computer systems have not been designed with security in mind. A good networking system must provide many elements of security that are otherwise not found in the PC environment. Any network that does not protect from unwanted snooping is not likely to be acceptable to the users. Thus, almost every networked system provides various levels of security, which may range from simple permission limitation (e.g., no access, read only, read and write) to elaborate systems designed to rival those of Fort Knox.

Once the LAN is integrated into the major data highways, users can have access to the vast array of public domain software, bulletin boards, and the like. Software vendors are also increasingly plugging into these networks, and debugs and application notes are often posted on the bulletin boards. The user groups provide a plethora of advice and information on the vast majority of popular packages.

6.4.3 Popular Network Operating Systems and Tools

Over the years a vast array of network operating systems (NOSs) have appeared, and most have disappeared equally fast. By far the dominant NOS for client-server computing has been Novell Netware™. Currently, three generations of Netware are in use, the choice depending on the requirements. Version 2.xx is the lowest (least costly) with 4.xx the top end of the range. Version 2.xx caters to 80286-based servers, while the higher versions take advantage of the 80386 and 80486 CPUs. They do use up a large amount of system memory on the server (2.5 MB on 2.xx, 4 MB on 3.xx, and 6 MB on 4.xx). In installations in which the server may be running Unix, Sun Systems' PCNFS™ is popular.

In peer-to-peer networking LANtastic™ is the dominant package. Other popular packages in this area are Novell Lite™ and LanSmart™. Microsoft Windows for WorkGroups™ (WFWG), and MicroSoft LAN Manager™ (now integrated into Windows-NT as the NT Advanced Server NTAS). As is to be expected, most of the peer-to-peer LANs also support ARCnet and Ethernet. Most of these networks can coexist inside a larger Novell environment and provide similar facilities. As is only to be expected, the system overheads are small (typically about 50–60 kB per node). As in a peer-to-peer environment, one can expect server nodes to be switched on and off; most packages check periodically and keep themselves posted about the current status of the system. Many of the packages also have a client-server mode, where one or more nodes can define themselves as servers (thus speeding up access for others) or clients (reduces system overheads) for better performance. The security and internetwork connectivity features are limited in many of these packages.

As many NOS systems do not support internetwork connectivity in a big way, a vast array of add ons for enhancing the capabilities of the NOS systems are available. These include mail, facsimile transmission and reception, TCP/IP connectivity, Unix tools for DOS, and so on. Taking advantage of the OSI architecture, most packages support a large selection of NOS shells.

6.4.4 Hardware of Ethernet Configurations

Ethernet is by far the most popular LAN for office use today. As has been discussed earlier, Ethernet can operate at up to 10 Mbps over distances of up to 2 km. Ethernet can be implemented using three kinds of cabling: thick, thin, and twisted pair. In this section the three types of cabling are discussed in some detail.

Thick Ethernet

Thick Ethernet, also known as 10 Base 5, is used for the backbone of large networks. It is seldom found in small office networks or in machine-to-machine linkages.

Thin Ethernet

Thin Ethernet, also known as Cheapernet and 10 Base 2, is the workhorse of the Ethernet working environment. Larger networks are often built by connecting segments of Thin Ethernet systems. The salient specifications of the 10 Base 2 system are as follows:

Type of cable:	RG58A (or B) /U (50 Ω coaxial)
Topology:	Daisy Chain
Maximum Number of Nodes per Segment:	30
Maximum Cable Length per Segment:	185 m
Minimum Cable Length between Nodes:	0.5 m

It should be noted that the end segments of the network must be terminated with 50-Ω resistors. The cable is connected to individual computers through BNC connectors and tees.

The main disadvantage of Thin Ethernet is that any failure means the entire network will go down. The most common causes of failure are cable failure and users inadvertently disconnecting and thus breaking the daisy chain; as a disconnected end is no longer terminated, the reflected waves from the open end lead to a network collapse. Up to three Thin Ethernet segments can be connected together through bridges.

Twisted Pair

The twisted pair or 10 Base T is the latest manifestation of the Ethernet standard. In a 10 Base T system, individual computers are connected to concentrators or hubs through ordinary twisted pair cable. The biggest advantage of 10 Base T is its robustness. In a daisy chain configuration, any breakage or failure leads to a failure of the entire network, which is avoided in a 10 Base T system. Individual 10 Base T hubs, each of which typically services 8 or 10 nodes, can be interconnected using either Thick or Thin Ethernet cabling. Thus, the hubs are daisy chained as in coaxial cable-based systems, with the nodes in a star configuration. This also makes moving machines easy since it does not disrupt the network. The cable limitation for the twisted pair is 100 m.

6.5 Networking in the Control Environment

In laboratory or pilot plant-level applications, some of the networking protocols discussed above are not cost-effective. In the case of distributed control, this may apply for the lowest level connections between the local computers and their slave controllers. The nature of information transfer here is of a few bytes rather than larger packets. The commonly used protocols are GPIB (IEEE488), CAMAC (used in particle physics research), RS-232C and RS-422. Many suppliers of process control equipment also provide proprietary interfaces, which are connected to the host computer through a gateway connected on the RS-232C port. This is done since most small computers do not have a large number of RS-232C ports available, and, therefore, some form of multiplexing is necessary. Also, the data rates are usually low enough to make this multiplexing desirable for cost considerations. The cost and cable length limitations of GPIB limit it to dedicated laboratory systems only. For industrial controllers, some form of low-cost serial communication is required. One popular way is to put a large number of controllers in parallel on an RS-422 line. It is then possible to control all these controllers in a fashion akin to that of GPIB, through a software protocol. In some cases, proprietary buses, working in a similar fashion, use a gateway that is either connected to a serial port or installed in the body of the main computer through a feature card.

Suggested Reading

Bose, S. K. 1985. *Digital Systems—From Gates to Microprocessors.* New Delhi: Wiley Eastern.

Bose, S. K. 1988. *Hardware & Software of Personal Computers.* New Delhi: Wiley Eastern.

Gallager, R. G., and Bertsekas, D. 1986. *Data Networks.* Englewood Cliffs, NJ: Prentice Hall.

Krutz, Ronald L. 1988. *Interfacing Techniques in Digital Design with Emphasis on Microprocessors.* New York, NY: John Wiley & Sons.

Ramos, E.; Schroeder, A.; and Simpson, E. 1992. *Data Communication and Networking Fundamentals using Novell Netware.* New York, NY: Maxwell Macmillan.

Tannenbaum, A. S. 1981. *Computer Networks.* Englewood Cliffs, NJ: Prentice Hall.

7
General-Purpose Interface Bus (IEEE 488 Interfacing)

Prem Malhotra

7.1 Introduction

Almost every R & D scientist today has come across the term General-Purpose Interface Bus at some time or the other. GPIB is also known as IEEE 488, IEC 625, and HPIB. For all practical purposes these terms can be used interchangeably. Most R & D equipment is available with the GPIB interface, attesting to its widespread popularity. In many cases it is built in, while in others it is an extra cost option. However, in practice the number of workers actually using the interface is much smaller. Having a rigidly standardized interface like GPIB offers immense benefits. Equipment from different manufacturers (all conforming to GPIB) can be interconnected to form very powerful systems with a minimum of fuss. In some of the more sophisticated instruments the full capabilities of the instrument cannot be exploited unless it is run under computer control. Furthermore, with the availability of excellent hardware and software for GPIB interfacing on PCs, there is no need to buy expensive dedicated computers for controlling GPIB instruments.

GPIB is essentially a set of passive lines used to interconnect testing and measuring equipment into complete instrumentation systems. The interface has been adopted by IEEE as a standard (Standard 488-1978, Digital Interface for Programmable Instrumentation). This chapter is essentially in two parts. The first part deals with the evolution, structure, implementation and applications of GPIB, while the second part describes one popular implementation of the GPIB interface for PCs.

7.2 Evolution of GPIB

Before the evolution of GPIB, the traditional approach to interface design in the field of bench-top instrumentation had been to provide each instrument with a number of control, data, and status signal lines. As the requirements of each instrument were different and every producer had his own approach to the problem, each instrument was unique, making interconnections of instruments a nightmare. While in a few rare cases it was possible to interconnect systems from the same source, mixing of instruments from different vendors was a Herculean task.

Though this approach gave a relatively simple and direct control of an instrument's programmable functions, it led to a variety of interface solutions. The net result was a dedicated interface structure for each instrument integrated into a system. Interface adapters were needed to accommodate different data formats, signal levels, logic conventions, protocols, timers, and so on.

The Hewlett Packard company evolved the interface bus HPIB (Hewlett Packard Interface Bus) in 1972 for the interconnection of instruments. It formed the basis for further work in standardizing instrument interfaces. Finally, in 1975, IEEE published its standard 488. Today, practically every manufacturer of scientific instruments supports GPIB, and most sophisticated test and measuring instruments have GPIB either as standard or as an option.

GPIB has been designed to satisfy the needs of most instrumentation systems. It permits the following:

1. Low-cost interconnection of up to 15 instruments over a limited distance (1 to 2 meters per instrument, up to 20 meters total).

2. Easy reconfiguration of a setup.

3. Interfacing of instruments with different data rates, varying from a few hundred bytes/sec to one megabyte/sec.

4. Interfacing of instruments with a wide range of capabilities.

5. Direct communication between instruments without having all messages/data routed through a control unit.

6. Modular implementation of the interface such that a subset of the interface suffices for simpler instruments.

7.3 Structure of GPIB

The GPIB consists of sixteen lines (Table 7.1), of which eight form the multiplexed data/command channel. Another three lines are dedicated to handshake, and the remaining five have management functions. A maximum of fifteen instruments can be simultaneously connected to one GPIB port. The bus length is typically 2 m/instrument, not exceeding 20 m in all. The data transfer rate can be as high as 3 MB/sec, but the transfer rates achieved in typical applications consisting of microprocessor-based instruments are of the order of 250 kB/sec. A GPIB cable is terminated (Fig. 7.1) at both ends by identical connectors (male-female back to back) that permit piggyback connections. Note that the piggyback connectors permit easy connection/disconnection of an instrument. Also, daisy chain and star connections can be freely mixed as convenient. The GPIB cables tend to be expensive due to the extensive shielding (of each signal line) and manufacture to very rigid standards. Figure 7.2 illustrates five GPIB instruments of different capabilities connected together on the same bus.

Table 7-1—The GPIB Lines

Line No.	Variable Name	Driving Device(s)	Function
1 to 8	DIO1 to DIO8	Talker, Active Controller	Data/Command channel. Data transmission between talker/listeners. Command transmission to all instruments.
9	DAV	Talker, Active Controller	Data Valid. Indication of validity of byte on data lines.
10	NRFD	Any Listener	Not Ready for Data. Indicates the non-availability of all listeners to accept new data.
11	NDAC	Any Listener	Not Data Accepted. Indicates the non-acceptance of the byte by all the listeners.
12	IFC	System Controller	Interface Clear. Initialization of GPIB Interface. Take over control of the system.
13	ATN	Active Controller	Attention. Multiplexing of data/command channel.
14	SRQ	Any Instrument	Service Request. Service request to active controller.
15	REN	System Controller	Remote Enable. Disabling the front panel controls.
16	EOI	Talker System Controller	End or Identify. Indication of end of block. Obtaining parallel poll response.

7.4 Operation of GPIB

In a configuration of instruments connected on GPIB, only one of them can be in the *Talker* mode at any instant of time, i.e., can drive the 8-data input-output lines (DIO1 to DIO8). There can be more than one *Listener*, i.e., instruments that would accept data from these eight data lines. If this were not so, there could be more than one talker of the line at the same time with the result being indecipherable gibberish. This is controlled through the handshaking lines.

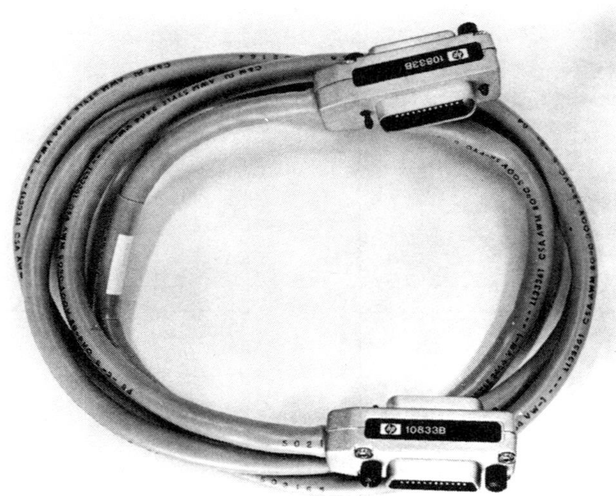

Figure 7.1—GPIB Cable (Note Double Ended Connectors)

(Courtesy Hewlett Packard Co.)

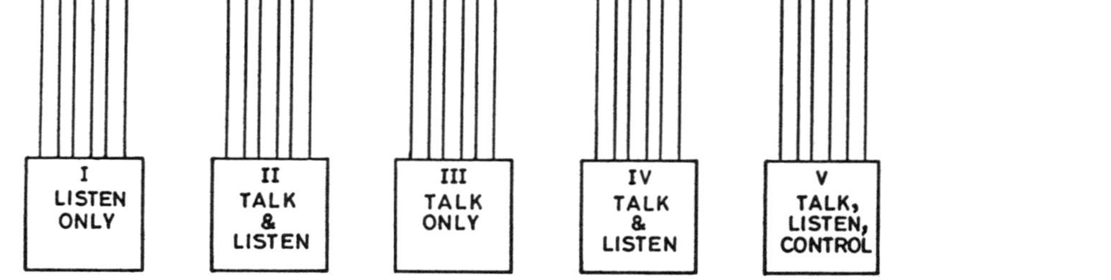

Figure 7.2—Five GPIB Instruments Connected Together

Transfer of each data byte on the DIO signal lines is accomplished via the set of three signal lines: DAV (data valid), NRFD (not ready for data), and NDAC (not data accepted). These lines operate in the interlocked handshake mode. The lines NRFD and NDAC are each connected in a logical AND (wire OR using open collector gates) to all instruments on the interface. The current talker controls the DAV line, while the potential listeners control the NRFD and NDAC lines. Figure 7.3 illustrates the three-wire handshake. All signals are active low. The data transfer proceeds as follows:

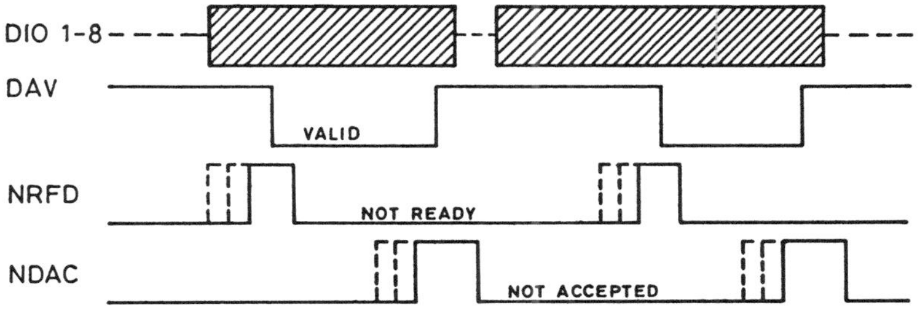

Figure 7.3—Three-Wire Handshake

1. Talker waits for all potential listeners to signify "ready for data" via NRFD, i.e., when it goes high. This line will not go high if any listener is not ready.

2. Talker places data on DIO lines and stabilizes it.

3. Talker asserts "data valid" on DAV (drives it low) after it has stabilized data on DIO lines.

4. Upon receiving "data valid," the potential listeners remove "ready for data" and assert "data accepted" on NDAC.

5. When all listeners have accepted data, i.e., NDAC is high, the talker removes "data valid" (DAV is pulled high).

6. Steps 1 to 5 are repeated for the next byte.

Data is transferred asynchronously at the rate of the slowest listener participating in the conversation. High-speed data transfer between two instruments will not be slowed down by a slower instrument on the bus, which is not an active listener during the conversation.

To effect a change in the talker/listener status of the instruments on the GPIB, one of these instruments is configured as a controller. In addition to being a talker or listener, it can transmit BPIB commands on the DIO lines and also drive other special lines such as IFC (interface clear), REN (remote enable) and EOI (end or identify). It also senses the wired OR service request (SRQ) line, which is used by devices on GPIB to draw the attention of the controller. In brief, the controller executes the following functions:

1. *Interface clear:* Initializes GPIB in all instruments and takes over as controller of the system. This is done by driving IFC low. In case there are multiple instruments on GPIB capable of being controllers, only the one designated as system controller can drive this line.

2. *Configure talker and listeners:* With the ATN line active, talk address of the instrument to be made the talker is sent through DIO lines. Listeners are enabled in a similar fashion. "Untalk" and "Unlisten" commands are used to remove the talker and all listeners from the bus.

3. *Attend the service request from other instruments on the bus:* An active SRQ indicates that one or more instruments are requesting attention. Controller interrogates each instrument one by one (called serial poll) or up to eight instruments together (called parallel poll) to determine the instrument requesting attention.

4. *Transfer GPIB control to another controller on the bus:* Even though there maybe more than one potential controller on the bus, at any given time only one of them can be the active controller. This function is necessary to hand over control to another controller.

5. *Monitor data traffic on DIO lines and interrupt communication on certain predefined conditions:* The controller acts as a passive listener and wakes up to take over the bus when the need arises. This may happen when a certain process reaches completion or an error condition occurs, etc.

6. *Enable or disable front panel controls on an instrument:* Many instruments allow the local lockout (LLO), under which the front panel controls are disabled during GPIB operation. The controller can disable and enable them to allow the instrument to be operated manually when not required by the controller.

Table 7.1 summarizes the functions of all GPIB signal lines.

It is possible to operate instruments on GPIB without a controller if the talker is to remain unchanged. This mode of operation was designed primarily for connecting a measuring instrument to a storage or hard copy device. However, with the much lower cost of computers today, this type of operation is very rare. In typical applications, the data gathered from some instruments (such as a digital multimeter) on GPIB usually needs some processing before being sent to other instruments (for instance, plotters). Therefore, a microcomputer is almost invariably used in a GPIB setup, and this normally has the GPIB controller function incorporated in it as well. This computer may have GPIB as an intrinsic part of it (e.g., some H.P. microcomputers), or GPIB may be implemented through an add-on system.

7.5 Implementation of GPIB

A GPIB interface is incorporated in an instrument not as an integral part of the functional instrument (referred to as device) but as a separate block. This allows the interface to be designed independently of the device specifications, and the same interface can be easily adapted to other devices. Figure 7.4 is the schematic of a GPIB-compatible instrument, which consists of the device and GPIB interface blocks. The device-dependent messages are routed through the interface and are treated as data on GPIB. Local messages between the device and the interface facilitate the transfer of device-dependent messages on GPIB (i.e., an interrupt from the interface to the device may indicate that a device-dependent message is waiting for it with the interface). Besides this, through the local messages the interface exercises certain controls on the device and vice versa, as permitted by the GPIB specifications (e.g. device clear, device trigger, etc.).

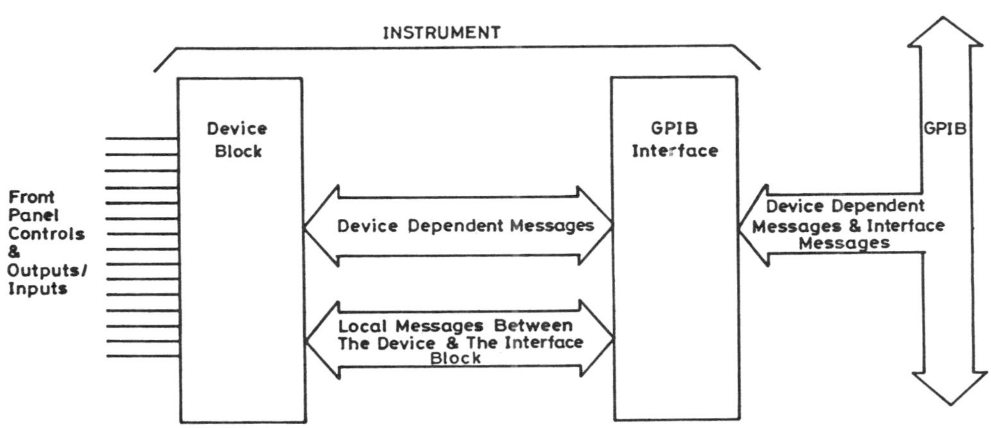

Figure 7.4—Schematic of a GPIB Instrument

The GPIB interface has been defined in terms of ten smaller functional units (called interface functions) doing limited jobs on the bus. Table 7.2 shows these interface functions. Such a modular approach to the interface permits building GPIB interfaces that have a subset of these functions, i.e., only the ones that are needed for the instrument. For example, a simple plotter may have only the acceptor handshake and listener functions implemented in its GPIB interface. The interface functions have been designed so that instruments with different subsets of the interface functions implemented can function together in an integrated system without any operational problems.

Table 7-2—GPIB Interface Functions

Mneumonic	Description	Function
SH	Source Handshake	Handles handshake for the talker/ handshaker during transmission (controls the DAV line)
T	Talker	Enables the instrument to send status/ data bytes when addressed to talk
L	Listener	Enables the instrument to receive data when addressed to listen
SR	Service Request	Sends service request to the controller, on SRQ line, on receiving the local service request message
RL	Remote/Local	Disables/enables the local controls using the REN line
PP	Parallel Poll	Enables the instrument to present one status bit to the controller
DC	Device Clear	Resets the device in the instrument
DT	Device Trigger	Triggers the device in the instrument
C	Controller	Enables the instrument to control the interface in all the instruments in the bus

The popularity of GPIB led to the design of large scale integrated (LSI) circuits to implement some or all of the interface functions in a single chip. Typically, 40-pin IC GPIB controllers are available from several manufacturers such as Intel, Motorola, National Semiconductors, and others. These are available both for instruments with and without microprocessors. In the first case, GPIB controllers connect to the local processor as any other peripheral IC, while in the latter case, the controllers are stand-alone and are pin programmable. Thus, it is fairly simple and inexpensive for manufacturers to provide GPIB in their instruments.

7.6 Applications

Several types of instruments are available with the GPIB interface; the following are a partial list.

1. *Measuring Instruments:* network analyzer, spectrum analyzer, waveform analyzer, digital counter, frequency counter, timer, programmable filter, power meter, digital multimeter, D/A converter, oscilloscope, etc.

2. *Stimulus Equipment:* function generator, voltage and current source, frequency synthesizer, waveform generator, etc.

3. *Display Equipment:* dot matrix printer, plotter, etc.

4. *Storage Equipment:* analog instrumentation recorder, magnetic tape drive, floppy drive, etc.

5. *Controllers:* programmable calculator, microcomputer, etc.

The main uses of GPIB-compatible equipment can be categorized as:

1. data logging;

2. automatic test jigs on production line;

3. process control; and

4. laboratory testing and measuring jigs.

Figure 7.5 illustrates a typical test jig consisting of a controller and stimulus and measuring instruments. This set-up may be used for simple current-voltage characteristics types of measurements. Another example of a test system may be one to test the frequency response of an instrumentation tape recorder. A frequency synthesizer would be the stimulus device feeding into the recording input of the tape recorder; the output from the playback circuit of the recorder would be measured by a digital voltmeter capable of making ac measurements. A GPIB-compatible plotter can be used to get frequency response plots. With a microcomputer-based GPIB controller, a program can be developed to automatically get a frequency response plot as soon as a new recorder is connected to the jig and a key pressed on the computer console. In this manner, even an untrained person can handle product testing on the assembly line.

7.7 GPIB Interface on the PC

This section describes the GPIB hardware and software interface for PCs called GPIB-PC2, manufactured by National Instruments (see Fig. 7.6). This family of interfaces is supported by many major manufacturers like IBM and Princeton Applied Research. The approach followed in this widely accepted card is described

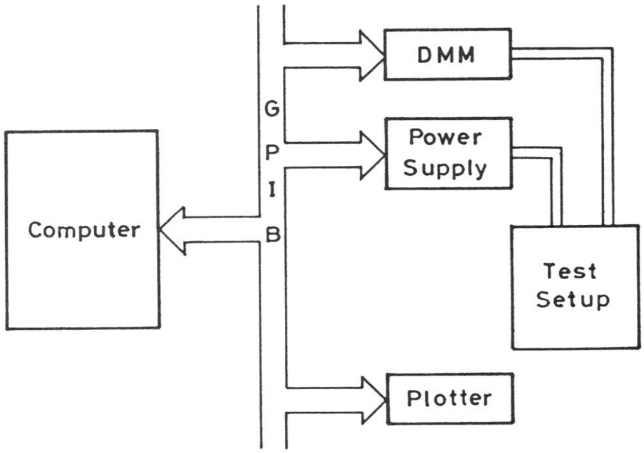

Figure 7.5—Typical GPIB-Based Test System

here. Cards from other manufacturers by and large follow a similar approach, though the exact terminology and syntax will differ.

The hardware consists of a half length card with a GPIB connector, which is inserted into a slot on the PC. The card is designed around a 40-pin LSI [M]PD7210 that supports all GPIB interface functions including the controller function.

The software consists of a MSDOS device driver, called DOS handler by the manufacturer, and a language interface for the DOS handler. Using this approach, the device driver remains the same, with a separate language interface for each language. This approach simplifies the development of interfaces for different languages but is slower than having a custom interface for each language. Here, the user first calls the language interface, which in turn calls the DOS handler. If there was a custom interface for each language, then the process would have been a single step rather than the two-step process of this approach. However, the loss of speed is not of any great importance since direct memory access (DMA) and other techniques are also available for cases where speed is of paramount importance. Such situations are very rare. This approach has the added benefit that all language interfaces behave in a similar fashion, and it is very easy to develop for different languages.

The DOS handler is a program (device driver) called GPIB.COM, which is loaded into the system memory from the secondary storage device (floppy or hard disk) at the time of the system boot through the CONFIG.SYS file. It interacts with the GPIB-PC2 as well as with the BIOS (in ROM) software of the PC and effectively

Figure 7.6—GPIB Interface Card

(Courtesy National Instruments)

masks from the user the lowest level board programming details, interrupt, and DMA handling. It permits the user to make use of a variety of handler function calls to interact with other devices on GPIB as well as the controller card. These function calls are divided into six categories according to their level of complexity and corresponding ease of use. The function calls can be used through several languages by linking a corresponding manufacturer-supplied code patch to the user program before execution. In addition, the user program must have external references to labels used in the patch, as specified by the manufacturer. The languages supported include assembler, BASIC interpreters and compilers, several versions of Pascal and FORTRAN, and packages like ASYST, Measure, etc. As the device driver module of the handler is common to all the languages, they all operate in the same fashion.

The language interface or code patch linked to the user programs executes the calls from the DOS handler area. For programs written in interpreted BASIC, the file BIB.M is the code patch that is added to the user program before execution; the file DECL.BAS contains the initialization code that is placed at the beginning of the user program source file.

The software package also contains a program called IBIC, which allows the user to execute function calls interactively from the keyboard, one at a time, with the status being displayed on the monitor. It is useful for becoming familiar with the GPIB functions as well as new GPIB equipment before writing complete application programs. It is a common practice to first test the operation sequence through IBIC before writing the actual procedures in the program.

The devices to be controlled on GPIB are installed through a program called IBCONF.EXE. The program modifies the file GPIB.COM to configure it exactly to the characteristics of the instruments that are to be controlled. The configuration provides details regarding the implementation and protocol requirements, which are not fully standardized. Perusal of the instrument manuals readily provides this information. It also allows the user to name the devices in a more convenient manner, e.g., it is more convenient to work with PLOTTER, DVM, POWER SUPPLY, IMPANAL, etc., than the cryptic dev1:, dev4:, etc.

In this chapter, BASIC with its implicit type declarations is used to illustrate the various aspects. Those unfamiliar with BASIC should note that the suffix % indicates an integer variable and $, a string variable. The following program to send data to a plotter connected on GPIB illustrates some of these function calls:

```
100 BDNAME$="PLOTTER"
120 CALL IBFIND(BDNAME$,PLT%)
130 WRT$="PD"
140 CALL IBWRT(PLT%,WRT$)
```

Statement 100 initializes BDNAME$ to the device name. IBFIND is the function call to open the device. It returns an integer PLT%, which is used as an identifier of this device later on. WRT$ is the string variable that is initialized to the command to lower the pen. IBWRT call sends WRT$ to the plotter.

All function calls return three parameters that indicate GPIB-PC2 card status, previous command execution status, and a count of the number of bytes actually transferred. These parameters are as follows:

IBSTA%	contains 14 bits of error information
IBERR%	contains error code if error bit in IBSTA was set
IBCNT%	contains byte count

The functions can be divided into the following two broad categories:

1. *Board Functions:* Such functions operate on the GPIB interface boards in the PC (a maximum of two boards can be installed simultaneously). These are low-level functions in that they perform specific rudimentary operations.

2. *Device Functions:* These are directed to a GPIB instrument connected to the interface boards in the computer, such as DMM, plotter, etc. These are high-level functions because they automatically perform a combination of board functions. Device functions are easier to use because many of the GPIB protocol details need not be understood.

7.7.1 Some Basic Device Functions

1. IBFIND (BDNAME$,BD%): Opens a device, BDNAME$ (an instrument or another GPIB board in the computer), and returns a number, BD%, called the descriptor, which can be used in further calls related to that device. It initializes the device in the software.

2. IBWRT (BD%,WRT$): Writes data from string WRT$ to the instrument specified by the descriptor BD%. It configures the instrument to listen and after the transfer of the data, sends the unlisten command also.

3. IBRD (BD%,RD$): Reads characters from the instrument specified by the descriptor BD to the string buffer RD$. Data is read until RD$ gets full or the end of string character (X in the case of some DMMs) is detected or the end message is received (EOS line goes active along with the last byte received). It configures the system to talk and, after the data transfer, to untalk.

4. IBLOC (BD%): All device functions automatically place the specified instrument in remote mode (by sending GPIB command REN). This call moves the instrument from remote to the local mode (in which front panel controls are enabled), until the next device function is executed by this instrument.

5. IBCLR (BD%): The instrument specified by the descriptor BD% is sent a GPIB message: selected device clear (SDC). It functions like a reset to the instrument.

6. IBTRG (BD%): It addresses and triggers the specified device. The GPIB message group execute trigger (GET) is sent to the instrument. It is used to trigger an event in the instrument such as collection of data in its local memory.

7. IBRSP (BD%,SPR%): The specified instrument is serially polled and the status byte returned is stored in SPR%. Its bit 6 is set if the device is requesting service.

8. IBONL (BD%,V%): The specified instrument is set on line if V% is non-zero; otherwise, it is set off line. For a device that is already on line, a call IBONL with

V% non-zero has the effect of device clear, i.e., all configuration settings are set to their default values. IBFIND automatically does IBONL with a non-zero V%.

The eight calls described above are needed for most applications involving the control of instruments under GPIB.

All function calls return three status parameters, viz., IBSTA%, IBERR%, and IBCNT%. Bit 15 of IBSTA%, if set to 1, indicates that an error occurred during the execution of that call. IBERR% contains the error codes. IBCNT% contains the actual byte transfer count if read/write operations were done on GPIB. It is advisable to check IBSTA % for the error condition bit after each function call in the user program.

7.7.2 IBCONF

This program is used to interactively modify the device driver file GPIB.COM. GPIB.COM is the DOS device driver used by all the language interfaces. The signal specifications and the basic functions in GPIB are rigidly defined, but many of the finer details (like EOS byte, 7-or 8-bit words) have been defined rather vaguely to permit a wide variety of instruments to use GPIB. These parameters are set in the handler through IBIC. It also allows instruments to be given names, which is very convenient. For example, in one system a DMM (digital multimeter) be configured at address 5, while in another it is at address 7. As long as the correct GPIB.COM is used and the instrument is internally configured to address 5 or 7 as appropriate, if it is called DMM in GPIB.COM, then an IBFIND call for DMM will set the correct address for both the cases. It should be kept in mind that GPIB.COM is loaded as a DOS device driver at the time of boot up. Thus, it is essential that the system be rebooted (warm boot, i.e., Ctrl-Alt-Del will do) after GPIB.COM is changed, in order to use the new configuration.

7.7.3 IBIC

IBIC.EXE is an interactive program used to execute the function calls one at a time from the keyboard. The instrument commands can be tested using IBIC before incorporating them in the user programs. This makes the development and testing of complex routines very easy, since they can be executed interactively, one at a time, through IBIC.

It should be borne in mind that in GPIB, while the connections and communications are specified, the meanings of the individual messages are not. A multimeter may use V for setting the voltage range, while a level controller may use the same character to mean volume. Thus, the exact content of the message is totally device-dependent. At first sight this may appear to be a weakness of GPIB, but in fact it is exactly this flexibility that permits very diverse instruments to use GPIB. It will be found that instruments of the same type from the same manufacturer have certain similarities in their syntax. When working with GPIB, good string handling is essential, since all conversations with the instruments are in the form of strings. For example, a multimeter may require the command string to include codes for mode, range, measurement time, etc. The response may contain the current mode, range, and value as well as other parameters. It will be for the user's software to generate and interpret these strings.

Suggested Reading

IEEE Standard 488. 1978. New York: IEEE Standards Office.

Low Level Measurements. 1984. Cleveland: Keithley Instruments Inc.

ISA Proceedings of 22nd International Instrumentation Symposium, Fundamentals of Test Measurement. 1976. Research Triangle Park, NC: Instrument Society of America.

8
Interfacing: Digital and Analog

Sanjay Gupta

8.1 Introduction

In order to use the digital computer for any application other than data processing, it has to be connected to some instruments, sensors, and actuators. This connection is called the *interface*. The real world is predominantly analog, not digital. Thus, it is necessary to have some device that can translate these analog signals to digital form to be intelligible to the computer and vice versa.

The interface may be unidirectional or bidirectional. Data logging requires a unidirectional connection since the computer is simply "listening" to the sensors. On the other hand, control requires a bidirectional connection since the computer is a "talker" as well as a "listener". It has to listen to the sensors and then take the appropriate corrective actions. This may, for example, be switching a heater on and off, or raising and lowering the stem of a control valve, or activating a pump. While switching a motor on and off is an example of a digital action, adjusting the position of a control valve is an analog movement. Thus, the control can be either digital or analog.

All buses used in computers are digital. Buses and communications were discussed in Chapter 6, while Chapter 7 was devoted to the discussion of GPIB, the General-Purpose Interface Bus. All these are examples of digital interfaces. In this chapter the interfacing of instruments that are not bus-based will be discussed. General-purpose laboratory interface cards typically offer eight or sixteen channels of analog input, two channels of analog output, and sixteen digital input-output lines. These boards are more than adequate for most laboratory applications. In this chapter the techniques for utilizing these resources will be discussed. Specialized boards are also available. For example, there are dedicated single-function boards for 32 or 64 analog inputs, 32 or 64 lines of digital input-output, 8 or 16 lines of analog output, stepper motor controllers, current loop drivers, and so on.

8.2 Digital Interfacing

Digital interfacing is seen much more often than is commonly thought. Many control situations require the activation of a control element, and, in most cases, simple on/off control is all that is required. Some examples are turning a pump on and off, switching a heater on and off, opening and closing of a valve, and switching a set of lines in a predetermined sequence to operate a stepper motor (see Section 8.5.5).

Situations involving digital inputs are somewhat less frequent in the laboratory. They often involve the monitoring of the status of control elements, or the sensing of certain alarms. For example, in a chemical reactor, there may be a need to take some emergency action if the pressure exceeds a certain threshold. In most such cases, a simple diaphragm switch is all that is required to actuate the alarm. This switch then gives the digital input to the controller. In the laboratory systems, alarm monitoring is less frequent, as more often than not the computer itself will be controlling it. However, in certain cases (specially in industry), these signals may also be required for safety or loss prevention considerations. There may be a valve whose closure is critical, and hence a micro-switch is attached to it, which must be sensed before any further action is taken. Another frequent requirement of digital inputs is in sensing the presence or absence of a fluid in a system.

Some older types of equipment do not have an interface but may have a binary output or, more commonly, a binary coded decimal (BCD) output. This can be used to hook up a computer for data logging. The ways in which this output are made available vary tremendously and can be discussed only in general terms. Each digit has four bits. BCD indicates a binary four-bit output, with counts above 9 disallowed; i.e., the bit patterns 1010 to 1111 that correspond to hexadecimal numbers A, B, C, D, and E are forbidden. The data can be read in and then interpreted suitably. In many cases some trigger or latch input may be required by the instrument, making the use of two parallel interface adapters (PIAs) (one in the input mode and the other in the output mode) mandatory.

While it is possible to use the parallel printer port for digital interfacing after suitable initialization, it is not very convenient. In general, digital interfacing is done through dedicated input/output ports, using PIAs. These are 8-bit devices; hence, the digital interfacing ports are in blocks of 8 lines. Each port corresponds to one byte of data that is simple and easy to understand. There are dedicated DIO cards with 2, 4, or 8 PIAs, giving 16, 32 or 64 lines on a single card. More commonly, what is found in the laboratory is a multifunction interface card with 16 digital I/O lines. To understand the handling of these ports, one requires insight into the way a PIA works. The typical laboratory interface card has 8 or 16 lines of analog inputs, two analog outputs, and two bidirectional PIA channels. Very often one PIA is used as an input and the other as an output channel.

8.2.1 Input and Output Ports

In the PC the digital interface is usually I/O mapped and not memory mapped. Memory and I/O mapping are discussed in detail in Chapter 9. The I/O devices can be of three kinds—input only, output only, and bidirectional. While the first two can support only unidirectional transfers, a bidirectional port is capable of being configured as either an input or an output port. Many interface cards (e.g., Keithley® System 570™, Adaptech® 20x series, etc.) use dedicated channels that are configured as either inputs or outputs. While bidirectional ports are more versatile, they do require a little more care in use.

On first sight it may appear that an output port is simply an input port reversed. This is not true, in general, due to the handshaking requirements. Digital interfaces can be of the handshake or non-handshake type. In a handshake system the availability or acceptance of data are signaled to the computer through a bit in one of the registers, as will be discussed shortly. In the case of non-handshake interfacing, there is no signal to the computer about the state of the other side.

A typical PIA interfaces to the computer through two adjoining addresses in the I/O map. Consider a bidirectional PIA. These registers are usually called command/status and data registers. The computer sends commands to the command register and reads the same register for monitoring the status of the PIA. The data register is used in either direction depending upon the mode in which the PIA is operating. It is called the data in or the data out register, as appropriate. The working of these registers is as follows:

When the computer wishes to transfer data out:

1. it sends the appropriate command to the command register to configure the PIA as an output device;

2. it sends the data byte to the data out register;

3. it monitors the status register to confirm that the data has been accepted (this is usually signaled by one of the bits);

4. once the data has been accepted, the cycle starts afresh.

When the computer wishes to read data in:

1. it sends the appropriate command to the command register to configure the PIA as an input device;

2. it checks the status register to confirm that the data is available; and

3. it reads the data from the data in register.

In the handshake mode of operation, the status register is monitored by the computer to ensure that no data is lost. In non-handshake operation the status register is not monitored. The computer assumes that the data has been accepted (for output) or is available (for input).

For faster data transfer the handshaking can be interrupt-driven. In the mode of operation discussed so far, the computer has to periodically read the status register to check whether it can proceed with the next byte of information. This involves loss of time. A faster way for doing the same job is to use an interrupt. In this case there is a dedicated line, called the interrupt line, from the PIA to the host processor that signals the acceptance or availability of the data. As soon as an interrupt is received, the processor enters the interrupt service routine to take appropriate action. While on one hand the interrupt is serviced faster, on the other the processor does not waste time polling the status byte periodically. This approach extracts its price in extra hardware.

As will be seen in Chapter 9, the PC does not allow too many interrupt lines; therefore, giving an interrupt line over to the PIA may not be possible. Interrupts are, however, used by the upper range interface cards, which often have fairly powerful onboard processors. These cards offer many high level commands, and they may use an interrupt line to get service from the host processor. This single interrupt line services all the features of the interface card. The interface card in this case is a small computer in its own right, which works under the control of the main computer, with a PIA working as the interface between the slave and the host. This situation is illustrated in Fig. 8.1.

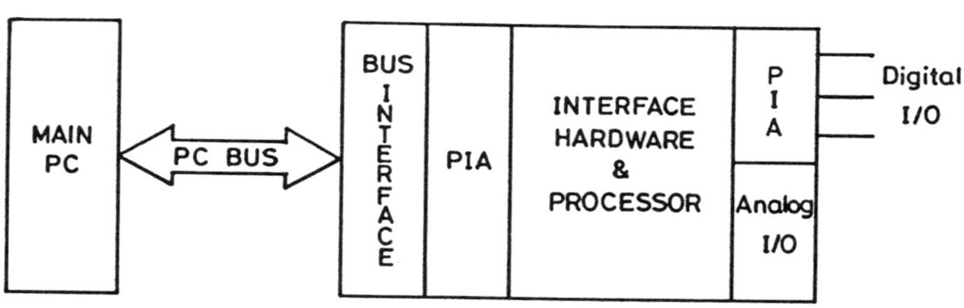

Figure 8.1—PIAs on an Intelligent Interface Card

Handshake or interrupt-driven interfaces are used by printers, cards, and devices directly interfaced with the computer bus. Thus, any interface card plugged into the computer bus ultimately interacts with the host computer through a parallel interface. On the other hand, process control systems connected to the computer very seldom use the handshake or interrupt-driven mode of communication; these devices may be relays, stepper motors, alarm inputs, among others, and are almost invariably handled in a polled fashion. This makes it necessary for the user to write his software to take care of the speed and other requirements of the connected output devices. In the case of input devices, the user must monitor or poll the input(s) periodically through his software.

8.2.2 Masking

Masking is of great advantage in using PIAs to control external devices. Many digital ports can be programmed with a mask. There are two ways of operating the mask. In one method the mask is logically ANDed to the output byte, and the result is what goes through (Fig. 8.2). This has the effect of keeping the output of the bits that have been masked out always zero, irrespective of the input. Thus, a mask of FFh (255 decimal) will lead to the output echoing the input and a mask of 00 will cause all output bits to always be zero. In the second mode of operation the masked-out bits retain their previous state. Both methods have advantages and disadvantages.

MASK	DATA	OUTPUT
1	X	X
0	X	0
1	X	X
0	X	0
1	X	X
1	X	X
1	X	X
1	X	X
1	X	X

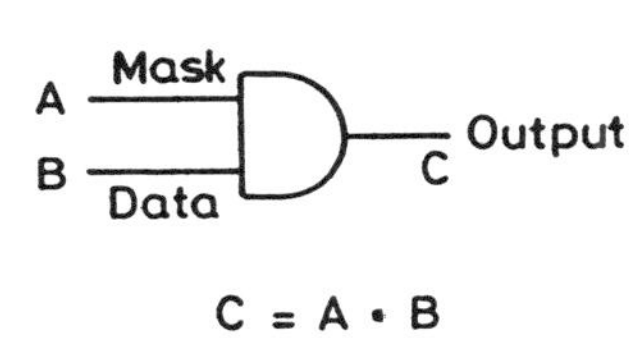

Figure 8.2—Effect of Masking on Digital Output

The facility of masking is of great use in real-life interfacing. There may be situations where more than one experiment share the same computer interface. Here, a proper use of the masks ensures that the program does not accidentally try to change bits that the current application does not use. This avoids the need to physically disconnect the lines going to the other systems. Also, in an application using many digital lines it may be desirable to temporarily disconnect some of them. This can be easily done by simply changing the mask without having to make any other changes in the program.

Let us take the case of a simple jacketed reactor. The flow of the heating medium is controlled by a needle valve actuated by a stepper motor. Bits 0, 1, 2, and 3 are used to drive the stepper motor. Bits 4, 5, 6, and 7 are used for the control of various inputs and outputs of the reactor. It may happen that the system becomes unstable, and the user needs to find out the cause. If masking was not available, major changes will have to be made to the software for testing. However, simply by masking out the lower four bits, i.e., using a mask of F0h, the stepper motor is disabled. Similarly, masks of EFh, DFh, BFh, and 7Fh can be used to disable bits 4, 5, 6, and 7, respectively. This makes testing and isolating the source of trouble very easy.

In the case of interface boards in which masking is not provided, it is fairly easy to implement in software, and the user will be well advised to route all digital outputs through such a routine. It may appear that one is unnecessarily adding a layer to the program, but in practice this pays for itself very quickly in much faster development and debugging. The use of a mask will ultimately simplify software in the case of larger applications.

8.2.3 Isolation

In many situations, isolation of the control and controlled signals are desirable or even necessary. For example, the digital output from the computer may be used to switch the power to a device on and off. The instrument at the other end may be a furnace or a pump or anything else. While it is quite possible to use the digital output to directly switch a thyristor or triac, it is not very desirable since a failure of the thyristor may, in certain cases, lead to feeding the mains to the computer, with disastrous results. This is avoided by using a solid-state (optically coupled) relay, or a magnetically coupled or reed relay circuit. Another situation may be an electrically noisy system where it may become necessary to isolate the controller and the controlled system to avoid feedback of noise into the controller. Again, solid-state relays are the preferred way to achieve this isolation; these are available in a vast variety.

8.3 Analog Interfacing

As was stated earlier, the real-world signals are almost invariably analog. Therefore, an analog-to-digital converter (ADC) is used to digitize an analog signal prior to its being fed to the computer. The digital output of a computer is converted to an analog form through a digital-to-analog converter (DAC). The DACs commonly used are of the R-2R network type.

Most commonly used ADCs use DACs internally. The cheapest (and slowest) ADCs are of the counter ramp type. The operation of a counter ramp ADC is illustrated in Fig. 8.3. A counter feeds a DAC, the output of which is compared with the input signal. As soon as the DAC signal exceeds the input signal, the comparator reverses and inhibits the counter. The counter then represents the digitized value of the input and is the first value larger than the input. The counter ramp ADC suffers from two major drawbacks. First, it is slow; a conversion of n bits will require about 2^n conversions. Second, the timing is not constant; a small signal will be digitized in only a few count cycles, while a large signal will take much longer, which makes precise timing difficult.

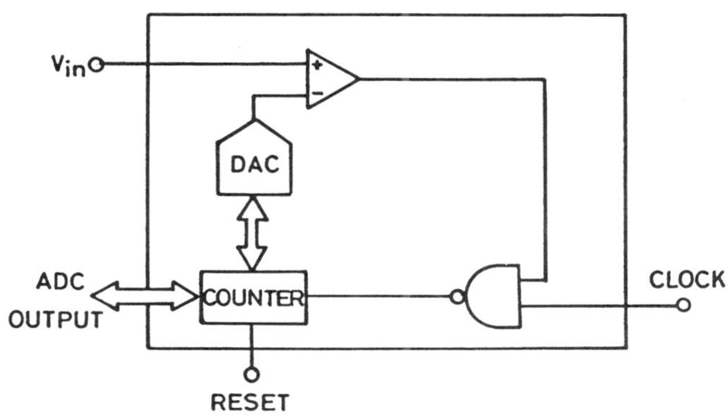

Figure 8.3—Schematic of a Counter Ramp ADC

Figure 8.4 shows a successive approximation ADC. The successive approximation algorithm relies on a binary search strategy to convert the signal. As is apparent, this also uses a DAC and a comparator, but the driving logic is different. First, the highest bit of the DAC is set to 1, with all other bits 0. The signal is then compared to the input. If the input is higher than the DAC output, then this bit is retained; otherwise it is set to 0. Next, the second highest bit is set to 1 with all the lower bits 0 and the output compared, and so on. Thus, for an n-bit converter, the final ADC value is available after n conversions. This value is the highest value that is less than the input signal. Thus, for the same signal, the output of a counter ramp ADC will be higher by one than that of the successive approximation ADC. The successive approximation ADC is faster and always takes the same number of cycles to convert a signal. The speed gain is very large. For example, for a 12-bit conversion, a counter ramp ADC may require as many as 2^{12} (4096) conversions, while a successive approximation ADC will require only 12. Due to the higher speed and constant conversion time, practically all the ADCs used in computer interfacing are of the successive approximation type. Counter ramp ADCs are very popular in low-cost multimeters, etc.

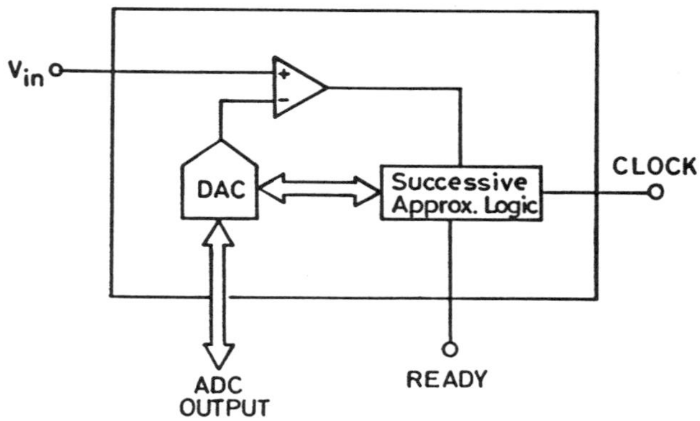

Figure 8.4—Schematic of a Successive Approximation ADC

8.3.1 Resolution

The first requirement in the selection of an ADC is the resolution, which is a measure of the precision with which the ADC can discriminate between almost equal signals. An n-bit ADC will provide a resolution step of $1/(2^n–1)$ of its range. As an example, consider the case of ADCs with a 10-volt input range. In an 8-bit ADC this will lead to a resolution of $10/(2^8–1)$ V, i.e., 39.2 mV. Ten-, twelve-, and sixteen-bit ADCs will give resolutions of 9.775 mV, 2.442 mV, and 0.1526 mV, respectively. Therefore, the ADC resolution should be selected depending upon the required accuracy of measurement (for most laboratory applications, a 12-bit ADC is adequate). Quite often the maximum voltage level available does not match the range of the ADC. In these cases, to achieve the full resolution of the ADC, a preamplifier becomes necessary. Many upper-range interface cards have built-in preamplifiers. Even then, the equivalent of 1 or 2 bits may be lost due to range mismatch. An increase in the resolution of an ADC leads to an increase in cost, so it may not be advisable to choose an ADC with a higher resolution than required. Even more important is the loss of throughput as the resolution is increased.

8.3.2 Throughput

The throughput of a system is the maximum rate at which it can convert signals. The main limitation in the development of very fast cards is the time taken by the output of the DAC to stabilize. In addition to the DAC, many interface cards may be further limited by the speed of the internal processor, the preamplifier, or the software interface. For example, a very popular range of laboratory interface cards is the DT2801™ series made by Data Translation. The DT2801 with on-board gains of 1, 2, 4, and 8 can do 12-bit analog-to-digital conversions at a maximum rate of 13,700 conversions per second. Its 16-bit version can only do a maximum of 2,500 samples per second. The DT2805™, which belongs to the same family (its on-board preamplifier has gains of 1, 10, 100, and 500) but can be used only in the differential input mode, can work only at 6,000 samples per second, with the 16-bit version slowing down to 100 samples per second (see Fig. 8.5). In practice the speed of 13,700 samples per second can be obtained from DT2801 by using either DMA transfers or very tight assembly language coding. Using the BASIC compiler, the speed drops to about 300 samples per second, and the BASIC interpreter slows it down by a factor of ten. This loss of speed due to the overheads of high-level languages are unavoidable unless DMA techniques or machine language programming are resorted to. It is axiomatic that 95% of the time is taken by about 5% of the code. As was discussed in Chapter 4, it may be possible to code this part in machine language and thus significantly reduce the time penalty.

8.3.3 Onboard Preamplifiers

Built-in preamplifiers make it easy to interface signals of widely varying ranges and exploit the resolution of the ADC to the full. There are three ways of setting the gain of the preamplifier. This may be done either through hardware or software or a combination of the two. While the data translation cards use software selectable gain, the Tecmar Labmaster™ and Advantech® PCL20x™ series prefer resistor selection; the Keithley System 570 uses a combination of the two.

Hardware selection normally involves the actual selection of the feedback resistor of the preamplifier, while in software selection the same is achieved through a software command. Hardware selection may be done by physically changing the resistor or by moving some jumpers or switches to select the appropriate gain. By appropriate choice of resistor, the gain may even be set to values not normally allowed by the supplier of the card. Hardware selection is simple but not very convenient. Software-selectable gain is much more convenient but not as flexible. Cards with software-selectable gain also tend to be more expensive due to the additional hardware required. In cases where the input signal can vary widely, software-selectable gain allows the best gain to be selected and used at all times, while this is not normally possible with hardware-selectable gain.

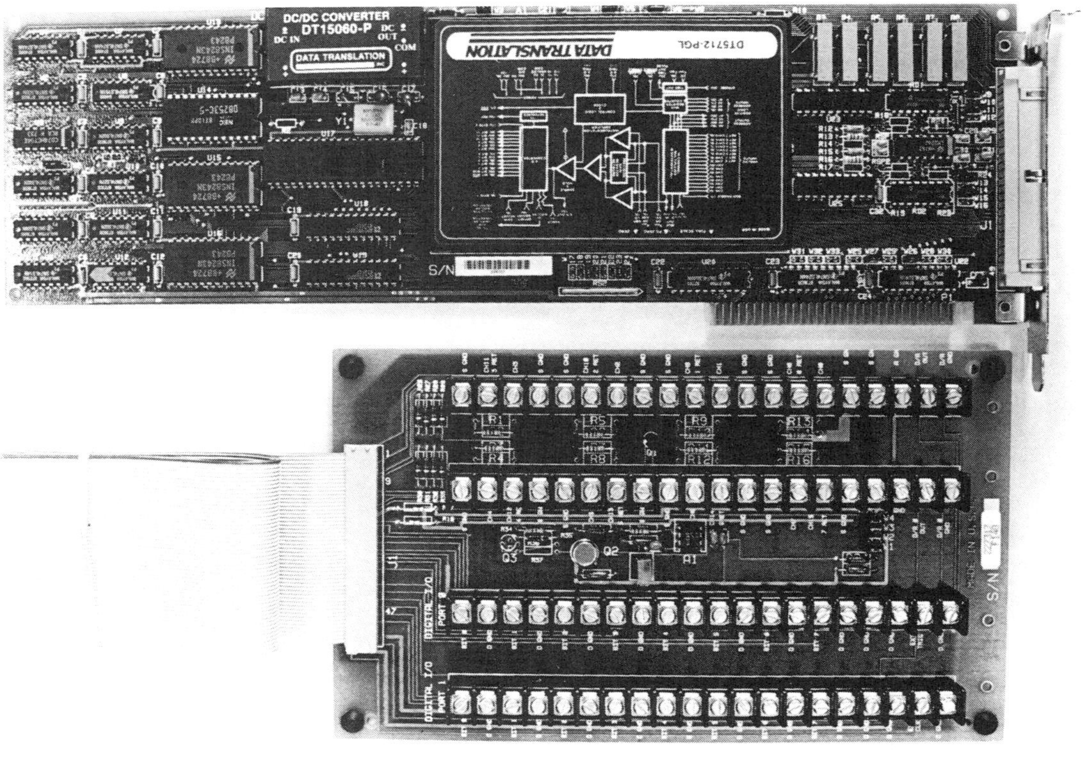

Figure 8.5—DT2805 Card and DT707T Screw Termination Panel

(Courtesy of Data Translation Inc.)

The combined hardware and software selection system provides for large (10x, 100x, etc.) changes in gain through hardware and smaller (2x, 5x, etc.) changes through software. This strategy is based upon the belief that, during the course of any single run, the signal is not likely to change over orders of magnitude, and the gain can then be optimized for the range of interest. However, for data logging in physical research where signals do change by orders of magnitude, this may not always be true.

Having an on-board preamplifier with selectable gains does not solve all the problems. Low-level signals can be seriously degraded by noise pickup in transmission. In such a situation a differential input to remove the common mode noise becomes necessary. Also, in cases of very precise measurements, long cable lengths, or a noisy working environment, it may become necessary to amplify the signal very close to the source in order to reduce the effect of noise pickup during transmission. Some of these considerations were discussed in Chapter 5.

8.3.4 Analog Data Value Concept

The analog-to-digital interface deals in digital quantities, i.e., the input to the DAC or the output from the ADC are in digital form. Interconversion of the digital and analog values is necessary for using the converters. The analog data value (ADV) is the digital value that corresponds to the analog voltage involved in this conversion.

In the case of digital-to-analog conversion, the DAC must generate the voltage required by the controller. The ADV is the number to be fed to the DAC in order to generate the required voltage. On the other hand, in analog-to digital conversion the output from the converter is the ADV, which then must be converted to voltage for subsequent use.

The specifications of the ADCs or the DACs in addition to the bits of resolution come in two forms. The more common practice is to specify the minimum (V_{min}) and maximum (V_{max}) voltages of the conversion range. In some cases the minimum voltage and the step size (V_{step}) may be specified instead. It is found that users often err in the interconversion of the ADV and the analog value (V). One should use additional checks to make sure that the conversions are correct.

A converter of n-bits will have 2^n–1 steps. Thus, the analog valve range, $V_{max}-V_{min}$, when divided into the number of steps, gives the step size as:

$$V_{step} = (V_{max}-V_{min})/(2^n-1) \qquad [1]$$

Then, for a DAC where V has to be converted to ADV,

$$\text{ADV} = (V-V_{min})/V_{step}) \qquad [2]$$

and, for an ADC where ADV must be converted to V:

$$V = \text{ADV} \times V_{step} + V_{min} \qquad [3]$$

In the event of the ADC having a preamplifier (gain = G), both V_{max} and V_{min} get accordingly scaled. If $V^1{}_{max}$ and $V^1{}_{min}$ are the maximum and minimum voltages for unit gain, the voltage corresponding to the output of an n-bit ADC with a gain of G is given by

$$V = \{\text{ADV} * (V^1{}_{max} - V^1{}_{max})/(2^n-1) + V^1{}_{max}\} \times G \qquad [4]$$

In some 16-bit cards the coding may not follow the above system, with ADVs from 0000 to 32767 (7FFFh) representing voltages above midrange and –32768 (8000h) to –1 (FFFFh) representing voltages below midrange. This may look strange at first, but it is totally in line with the internal representation of integers in computers where integers are usually 16-bit with a 1 in bit 15 indicating a negative number. Here also, 0000 to 7FFFh are positive numbers from 0 to 32767 and 8000h to FFFFh are negative numbers from –32768 to –1, respectively. This can be easily accounted for by adding 32768 to the analog data values involved in the above equations, or by setting V_{min} to 0 in Eqns. 2 and 3.

A very common error is to assume that an n-bit converter has 2^n steps instead of 2^n–1 steps. This mistake is so common that some manufacturers of interface cards so adjust their preamplifiers that the nominal range actually corresponds to 2^n intervals. This well-intended step of the card manufacturers may further complicate matters for some of the users, and the specifications of the cards must be looked into with great care. For example, the 12-bit versions of the DT2801 for a gain of 1 have an actual working range from –10 V to 9.9951 V. The step size is 4.8828 mV. Thus, the nominal ±10 V range is applicable for 4096 (2^{12}) and not 4095 intervals.

8.4 Practical Aspects of Interfacing

This section discusses some common signal transducers and actuators and how they are used. Even though the signal sources and control elements differ widely, they can be categorized into similar classes as explained in the following.

8.4.1 Low-Level Signal Sources

These sources give signals in the millivolt range. Examples are thermocouples, platinum resistance thermometers, and strain gauges, among others. Due to the very low level of the signals, extreme care is necessary in handling them. For example, the sensitivity of a thermocouple is ~10 µV/°C. Furthermore, an electrically heated furnace may have a strong electromagnetic field at the mains frequency, which can introduce noise into the thermocouple. To successfully measure the thermocouple signal, it becomes necessary to use differential inputs. Even then, digital filtering by averaging a large number of values is often necessary. A typical temperature measurement may involve the following steps:

1. Take a large number of measurements of the thermocouple signal.
2. Sort these values and average after discarding the extreme values.
3. Repeat steps (1) and (2) with the room temperature sensor for cold junction compensation. For example, the Data Translation DT707T™ termination panel gives an output of 0.5 mV/°C.

4. Look up the thermocouple table to get the voltage that corresponds to the measured cold junction temperature.

5. Add this voltage to the voltage measured in (2).

6. Look up the thermocouple table to find the temperature that corresponds to the voltage obtained in (5).

In Chapter 12, Section 12.7 this is done in the procedure XMT where sixteen measurements are taken and the median eight retained after discarding the four extreme values at either end. Since chromel-alumel thermocouples are linear within three degrees up to 1000°C, the thermocouple look-up has been simplified to a linear calculation.

The signal quality can be improved significantly by adding a simple RC filter to the input. In electrically heated furnaces, mains and low-level voltages are in very close proximity, and some furnace liners have a tendency to becomeconductive at elevated temperatures. In addition, to protect the ADC input from possible high-voltage leakage, it is desirable to connect two anti-parallel silicon diodes in addition to the filter. These will limit any leakage to approximately 650 mV, which will not damage the ADC. Figure 8.6 shows the filter and diodes in place. It may also be desirable to connect the positive input of the thermocouple to a relatively high dc voltage (5 or 10 V) through a high-value resistor (tens of megohms). This voltage will normally be shorted by the thermocouple; but, in the case of a thermocouple break, the signal will go high and will be interpreted as an unacceptably high temperature. For cooling applications, the resistor may be connected to ground, leading to a down-range signal in case of a break in the thermocouple. A break, once sensed, can be used to signal an alarm condition. Some interface boards that use external preamplification (e.g., Advantech PCL20x series with PCL7701/2 preamplifiers) have provision for adding a capacitor to the feedback loop of the preamplifier. This provides an easy way to filter the input signal.

For strain gauges, etc., similar precautions must also be taken. In the case of platinum resistance thermometry, it is possible to pulse the excitation of the sensor and thus minimize the effects of sensor heating. This is illustrated in Chapter 12, where programs using both ADC, DAC and a DMM are given.

8.4.2 Normal Signals

Signals in the 1- to 10-volt range seldom require much additional conditioning and can be used directly by the ADC. These are very successfully handled through normal single-ended inputs unless the signal itself is floating with a common ground to the computer.

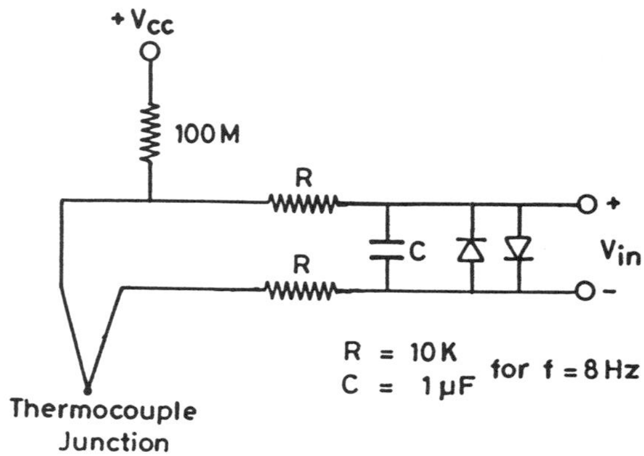

Figure 8.6—Filtering and Protection

8.4.3 High Impedance Signals

Signals from sensors such as a pH probe or an oxygen probe are characterized by a low-level output (~100 mV) and a very high source impedance (≳100 megohms). As the input impedance of the ADCs is typically in the 100-megohm range, a direct connection will result in loading of the sensor, leading to false readings. In these cases a standard pH meter or a redox meter may be necessary. The meter can, if required, be connected to the computer by tapping the display signal as explained in Section 10.4.

8.4.4 Industrial Process Control Signals

In the world of industrial process control, the most common output of sensors is a 4–20 mA current loop. Similarly, most actuators also work with a 4–20 mA control signal. The main advantage of a current loop operation is that the signal is not affected by long cable lengths. Also, this leads to standardized operation, making it simpler to develop interfaces. Quite often the signal converter that converts the transducer output to a linear 4–20 mA signal includes electrical isolation of the input from the output. Similarly, many analog control elements such as valves are operated by 4–20 mA signals through current-to-pressure (I/P) converters. Furthermore, a signal below the 4–mA level indicates either a break in the line or a sensor malfunction. A signal above the 20–mA level invariably indicates a system malfunction.

For A/D conversion, a shunt resistor (normally 250 ohms) is required at the ADC end to convert the current loop signal into a voltage. The signal of 4–20 mA then becomes a voltage signal of 1–5 volts, which is easy to handle with single-ended inputs. As long as the sensing system works on the 4–20 mA current loop, the actual quantity being measured is irrelevant. Signal conditioners for the most common sensors are commercially available. Many of these offer a current loop output or, alternatively, the voltage output can be converted to current through a 4–20 mA sensor transmitter such as the Analog Devices AD693™, which works with inputs of the millivolt level. This is illustrated in Fig. 8.7.

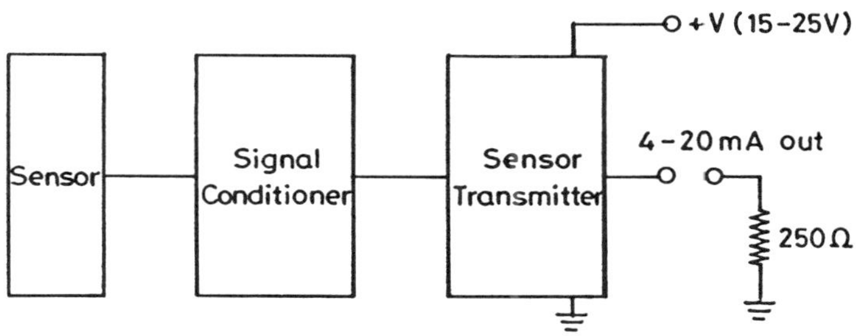

Figure 8.7—Signal Conditioning and Current Loop Operation

For D/A conversion, many DAC cards allow the user to select current loop operation. Such operation is almost invariably available on cards intended for industrial use. On the other hand, most cards intended for laboratory use do not allow such operation. Also, the currents that can be drawn from the DAC outputs are very small (about 2 mA). In these situations, if a current loop operation is desired, the best way is to use special ICs designed for driving the current loop from a volt input (typically 0–1 V, or 0–10 V). Some common devices for this are the Analog Devices 2B20™, 2B22™, and 2B23™ voltage-to-current converters. Except for the 2B20, the others also offer isolation of the input from the output. The connection of a DAC to a pneumatic valve is shown in Fig. 8.8.

8.4.5 Digital Inputs and Outputs

Interface cards invariably use digital input and output belonging to one of the standard digital logic families, the most common being TTL (usually LSTTL) or CMOS logics. For inputs to the computer, most sensing elements will conform to one of these logics, and inexpensive ICs are available to convert from one family to another. The connection of limit switches is very simple. In all of these are microswitches whose operation is to be sensed. The circuits given in Fig. 8.9 will work with most

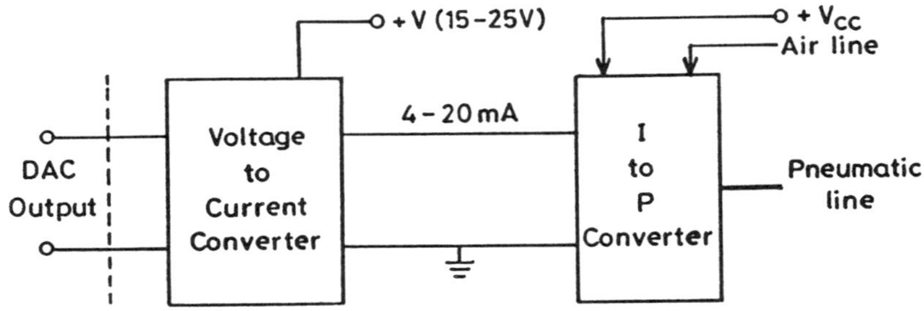

Figure 8.8—A Current Loop Driving a Pneumatic Valve through an I to P Converter

TTL and CMOS families. Figure 8.9a will give a logic 1 on operation, and Fig. 8.9b will give logic 0. While most cards in the input mode offer one TTL load (2 mA) or less, some cards with TTL outputs may support a fan-out of only 1. In these cases buffering may be necessary. An opto-coupled solid-state relay often requires 3 mA or more to operate, which will correspond to a fan-out of 2. However, in practice in more than 95% of the cases, the sensitivity is enough to use the *source* mode of the digital output, i.e., under the overlap of the digital output line. If strict compliance with the specifications of the various components is required, the connections discussed later (illustrated in Fig. 8.13) should be used.

Figure 8.9—Microswitch Connected To Give a Logic Signal (a) Logic 1, (b) Logic 0; When Closed

Many control situations do not require analog inputs. The control action is done by switching the input either on or off. Examples of these are heaters, solenoid valves, pumps, and so forth. Solid-state relays are the almost universal interface elements of choice in these cases. Optically coupled relays are replacing the other types (like reed relays) progressively. In these, the output of the computer is used to drive a light-emitting diode, the light from which, in turn, operates a photo-transistor or photo-thyristor. A vast range of such relays is available for most applications. A practical circuit for these is given in Fig. 8.10. This circuit is used to operate solenoids of 12 V, 1 A rating.

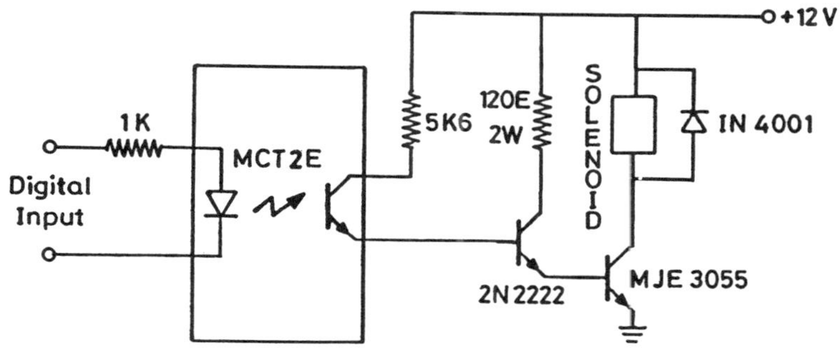

Figure 8.10—Solid-State Relay for 12 V 1A DC Operation

It is very unlikely that the relay required will not be available commercially. What may sometimes need to be added is simple protection for the relay. When inductive loads (solenoids, motors, etc.) are connected, a reverse emf is generated at the time of switching off. These voltages can be very high and may damage the switching transistor or thyristor if no precautions are taken. Figure 8.11a gives the typical protection circuitry for dc and Fig. 8.11b for ac loads. In the former, a diode of a rating equal to the device is connected anti-parallel to it, and, in the latter, an RC circuit is connected across the thyristor or triac.

Stepper motors are becoming increasingly popular as the controllers for valves and so on. They offer precise control of the rotation of a shaft through digital signals at a very low cost and are quite rugged. Pneumatic controllers are being progressively replaced by stepper motor-driven controllers. Mostly, four-pole motors are used. Let the poles be A, B, C, and D. The connections (see Fig. 8.12) are very like those for four solenoids. In many boards, while the digital output channel cannot source sufficient current to drive the relay, the sinking capacity is adequate. Under such circumstances the circuit in Fig. 8.13a can be used, keeping in mind that now the logic is negative, i.e., a LOW state corresponds to on and a HIGH state to off. In other words, the entire logic sequencing discussed so far has to be reversed. Alternatively, a simple buffer can be used to drive the solid-state relays (Fig. 8.13b). Figure 8.13c shows a variant of 8.13b in which the LED lights up when the relays are properly connected and energized. The 5-V (or 12-V) power is often available on the connector on the interface board itself.

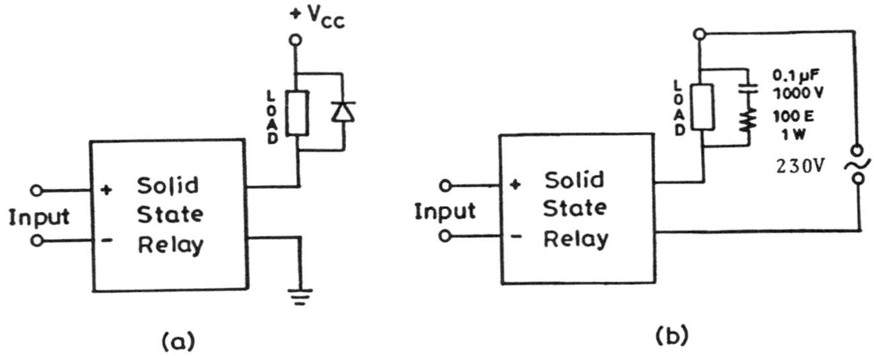

Figure 8.11—Protection of Solid-State Relays from (a) DC, and (b) AC Transients

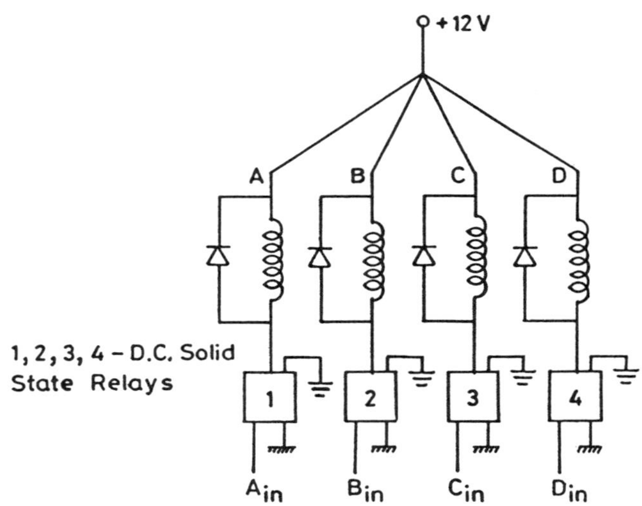

Figure 8.12—Digital Signals Driving a Stepper Motor

In normal or full-step operation, the motor is rotated by switching the lines in the sequence given in Table 8.1. The sequence is repeated as often as required. In the case of reverse rotation, the sequence is also reversed.

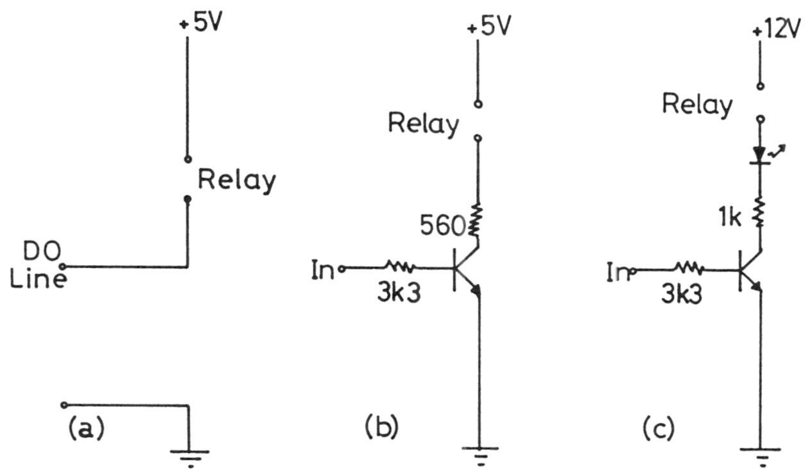

Figure 8.13—Driving a Solid-State Relay
(a) Sink Mode, (b) Buffered, and (c) Buffered with LED Indication

Table 8-1—Full-Step Operation of a Stepper Motor

Step	A	B	C	D	Hexadecimal Output
1	1	1	0	0	C
2	0	1	1	0	6
3	0	0	1	1	3
4	1	0	0	1	9

The motors can also be operated in half-step sequence. The only disadvantage of half-step operation is that the torque is lower due to only one pole being energized at alternate steps. The operation sequence in half-step mode is given in Table 8.2.

Table 8-2—Half-Step Operation of a Stepper Motor

Step	A	B	C	D	Hexadecimal Output
1	1	0	0	0	8
2	1	1	0	0	C
3	0	1	0	0	4
4	0	1	1	0	6
5	0	0	1	0	2
6	0	0	1	1	3
7	0	0	0	1	1
8	1	0	0	1	9

It will be seen that the alternate steps are the same as before, with additional steps between each of these steps. In the additional steps, three poles are off with only the pole energized that does not switch on the next step. Stepper motors are available with steps of 0.9, 1.8, 3, and 6 degrees corresponding to 400, 200, 120, and 60 steps per revolution. For speeds up to about 60 rpm, the simple circuit given in Fig. 8.12 is sufficient. For higher speeds, more complex chopper drives are necessary. Simple programs for driving stepper motors in the full-step mode are given in Turbo PASCAL as well as BASICA in Chapter 12.

Suggested Reading

Barney, G. C. 1985. *Intelligent Instrumentation: Microprocessor Applications in Measurement and Control*. London, UK: Prentice Hall Intl.

Control Engineering. October 1987, vol. 34, no. 10. Barrington, IL: Cahners Publishing.

Eggbrecht, L. C. 1983. *Interfacing to the IBM Personal Computer.* Indianapolis: Howard W. Sams.

Krutz, Ronald L. 1988. *Interfacing Techniques in Digital Design with Emphasis on Microprocessors*. New York, NY: John Wiley & Sons.

Morris, A. R. 1993. *Principles of Measurement and Instrumentation*. London, UK: Prentice Hall.

User Manual for the DT2801 Series. 1985. Marlborough, MA: Data Translation Inc.

9
Interfacing the PC Family

Sanjay Gupta

9.1 Introduction

The principles used in interfacing have been discussed in the earlier chapters. These principles are of a general nature and do not deal with the specifics steps required to actually introduce an interface card in a PC. This chapter discusses the specific factors required to successfully install an interface card in a PC. In order to carry out this operation successfully, some information about the hardware of both the card and the PC is obviously required. Even though the architecture of the PC is standardized, there are variations between different machines, depending upon the exact configuration being used. The selection of the address, DMA channel, and Interrupt channel will be determined by the exact hardware present in the system. Due to this, the better interface cards provide the user the facility to select the exact configuration for the particular card. The default configuration is normally so chosen as not to interfere with the hardware present in a vast majority of cases. Before selecting a particular card for use in the PC, the following aspects should be considered.

Suitability: Ensure that the specifications meet your requirements. Look into DOS, language, etc., that the particular card will require. It is always advisable to procure the software drivers with the cards. The manufacturer understands his cards the best. Later on, it may sometimes become necessary for a user to replace the manufacturer's driver by his or her own version, but such cases are extremely rare. Also, the experience gained by the extensive use of the card will make this task much simpler.

Documentation: The card selected should be supported by adequate information about its hardware and software requirements, in addition to information about its capabilities. Many low-cost cards are supplied with very little documentation. It is absolutely essential that the resource requirements, etc., are extremely well documented. Otherwise, the chances of running into serious troubles at some stage are very high. Hence, one should choose a card from a reputable manufacturer who has significant experience and wide acceptance.

Ease of Installation: Some cards now come with configuration utilities. These can often make installation a breeze. Some utilities will interrogate the system and then indicate the jumper settings and so on to use. Software configurable cards often come with installation software that automatically takes care of the entire setting up, driver installation, etc. However, for interfacing applications such hardware may not always be available or may be prohibitively expensive. On the other hand, software configurable cards with auto installation drivers are rapidly becoming *de rigueur* in the networking world.

9.2 Hardware and Software Requirements

Due to the very high degree of standardization in PCs, the hardware requirements for the successful addition of an interface card into the system are minimal. In the early days of the PC, when multifunction cards were not available, there were often problems of slot availability, power, etc. These problems have now almost vanished following widespread use of multifunction cards. What must be ensured is that a slot of the desired size is available. The following considerations apply:

Size: Ensure that the slot has sufficient length to accommodate the card in question. Many compact systems will not accept full-length interface cards.

Bus: There are two aspects in the bus selection, namely, electrical specifications and clock speed. The various interface standards in common use are XT or 8-bit bus, ISA or AT bus, EISA bus, VESA bus (ISA or EISA) and MCA. MCA is a separate category altogether and will not be discussed here. The XT or 8-bit card will work in all the other variants (physical limitations permitting). Similarly, the ISA card will work in EISA and VESA machines. An EISA card must have an EISA bus (with or without VL bus) and so on. The basic point to note is that ISA is a superset of the XT bus, and EISA, in turn, is a superset of ISA. VL bus is an extension on ISA or EISA.

As for clock speeds, the original PC and XT ran on the same speed as the CPU; i.e., 4.77 MHz. With the advent of the AT, a split clock system was introduced in which the CPU and bus clock speeds were not the same. Industry by mutual consent standardized on a 8-MHz bus speed. This has recently been increased to 10 MHz. Many popular 33 MHz systems, however, run the bus at 8.25 MHz (33/4 MHz). This leads to the possibility that an older card (designed for a maximum bus speed of 8 MHz or less) may not work with a more modern machine running the bus at 10 MHz. This problem may manifest itself as a total or a partial failure. Partial failures can sometimes be tricky to discover. The author had one analog input board designed for 8 MHz running happily in a machine with a 10 MHz bus, as long as one conversion at a time was being done. However, as soon as multiple conversions were requested, the machine would hang up after two or three measurements. In another case, the slower card prevented the machine from even booting up. However, quite often the bus can be slowed down by changing the setup (especially in machines with an AMI

BIOS). Alternatively, various public domain software utilities are available for achieving the same purpose.

Thus, it must be ensured that enough power is available. Modern designs require far less power than machines of old. However, some compact machines may not have enough power to drive power-hungry peripherals. Some application software may require a numeric coprocessor, a specific graphics card, etc. If so, these must be installed in the PC. It should be ensured that the correct version of DOS, languages, etc., are used. Many language interfaces for DOS drivers are specific to the version of the language being used, and this should be checked with care. Also, especially if using C language libraries, ensure that the correct memory model is being used. The user must as far as possible try to get the software drivers for whatever hardware is procured. The manufacturer knows his card best, and there is no point in spending long hours reinventing the wheel. Once sufficient experience has been gained, the user may attempt to write his or her own handler for some specialized applications (drivers are much more difficult to write and must almost invariably be written in machine language). BASIC users drivers written for BASICA (or GWBASIC) are not likely to work with QBASIC, which is shipped with DOS 5.0 onwards.

9.3 Flexibility

The interfacing hardware one buys should have the flexibility to allow one to select the I/O port address and select or disable DMA channels and/or interrupts. This will allow one to successfully interface the card into the available space and resources of the system. While the first card one buys can probably just be plugged in as supplied, the second card is likely to require some custom configuration in terms of address, interrupt, and DMA channel assignments. A little extra investment at this stage can avoid a lot of grief later. Fairly well established compatibility suites are available, and most reputable hardware and software on the market has been validated through these. Most manufacturers maintain lists of hardware and software that run on their systems. However, interface cards seldom figure on these. On the other hand, manufacturers of the interface cards may be able to advise on the compatibility of their cards with a particular system, especially if it is from a large manufacturer. For the inexperienced user, cards that can autoconfigure through software are a boon, as they make configuration as easy as plugging in the card and at most running the configuration software. Many other cards come with software that can interrogate the system and then advise on the settings to use.

Having established the basic norms, the various steps involved in interfacing a card are now discussed. One should carefully read the manual that accompanies the card. Then the user should proceed as follows:

1. Determine which slot will be most suitable. The card size, external and internal cabling needs, etc., have to be looked into for this. For example, a GPIB card should not be put in the outermost slot, since its cable connector may interfere with the chassis of the PC.

2. Having selected a slot for a card, proceed to select the I/O address, interrupts, and DMA channels for the card. The jumpers or DIP switches on the card should be then set. If possible, for preliminary testing, try not to change any of the factory settings. This may sometimes necessitate the removal of some other cards. Once proper operation has been verified, the settings may be changed and the other cards replaced. This approach is desirable since a departure from factory settings may require changes in the software, which may be difficult to identify in an unfamiliar manual. If a single card is being introduced into the PC, the chances are very high that the configuration set by the card manufacturer will work. For subsequent card(s) this probability rapidly decreases, and custom configuration becomes more and more necessary. This may require the disabling of interrupts on one card, DMA on another card, and so on. Great care must be taken at this stage since an attempt to share the same interrupt or DMA channel between two devices may even result in a burnt out IC.

9.4 Input-Output Addressing

The first parameter to be selected by the user is the input-output address to be assigned to the add-on card. The PC has a 16-bit I/O address field. Thus, in principle, there can be 65,536 I/O ports. In practice, the number is far fewer. On the boards only the lower 10 bits are decoded, which reduces the number of independent addresses to 1024 (2^{10}). Bit 9 was originally used to select between the motherboard (0) and slots (1). Thus 0000 to 01FF were available on the motherboard and 0200 to 03FF on the slots. However, in modern designs this distinction is only academic. The systems are as likely to have all-in motherboards as not. Some systems have all the peripheral support on the motherboard, with the CPU, etc., on a daughter board, and so on. Quite often the decoding is partial, which leads to one device occupying quite a few of these addresses. Many add-on cards also use partial decoding. The I/O map of a standard PC system (motherboard and add-on cards) is given in Table 9.1. Many of the assigned I/O addresses in the table will almost certainly be free on any system, since it is very unlikely that all the possible additional cards are present. However, the use of these assigned addresses should be avoided, since later upgrades may otherwise force the user to again reassign his or her special add-on cards. Reference to the documentation of the system should yield the exact map of a particular system. It is advisable to prepare an I/O map of the system for reference and update it whenever any additions or changes are made. This little care can save a lot of time and grief later.

Table 9-1—Input-Output Map of a Standard PC

0000 – 000F	DMA Chip (8237)
0020 – 0021	Interrupt Controller (8259)
0040 – 0043	Timer Counter (8253)
0060 – 0063	Parallel Interface Chip (8255)
0080 – 0083	DMA Page Registers
00A0	NMI Mask
0200 – 020F	Game Controller
0210 – 0217	Expansion Unit (Seldom Used)
0278 – 027	LPT2
02F8 – 02FF	COM2
0320 – 032F	Hard Disk
0378 – 037F	LPT1
03B0 – 03BF	Monochrome & Printer Adapter
03D0 – 03DF	Color Graphics Adapter
03F0 – 03F7	Floppy Adapter Card
03F8 – 03FF	COM1

The PC supports 256 (0100H) interrupts. Each vector pointer requires 4 bytes. Thus, 1k of RAM from 00000 to 003FF is dedicated to the vector pointers. The *DOS Technical Reference Manual* gives full details of these.

It is also possible to map a device into the 20-bit (1 MB) memory space of the PC. The memory map of the classic IBM-PC is given in Table 9.2. In PC compatibles, which did not use ROM BASIC, the BIOS was either 8 kB or 16 kB in size starting at FE000 or FC000, respectively. Units with ROM BASIC almost invariably used an 8-kB BIOS, with ROM BASIC occupying 32 kB. In AT-class machines, the BIOS is either 64 kB or 128 kB. The address space on ISA machines is 24 bit, permitting the system to address up to 16 MB of memory. EISA and MCA (32-bit) systems can address the full 32-bit (4 GB) address space of 386 (and higher) processors.

Memory-mapped interface cards are not as common as I/O-mapped interface cards. It may be noted that the display on a PC is memory-mapped. The user is not likely to require detailed knowledge of memory mapping, since it is extremely unlikely that there will be a need to interface two memory-mapped cards at the same time, and the manufacturers of the cards are normally very careful to select a free area of memory. The odd card that uses memory mapping will almost invariably work on the default configuration.

Table 9-2—Memory Map

00000 – 9FFFF	RAM
A0000 – BFFFF	Reserved for Display Buffer (Graphics Cards)
B0000 – B3FFF	Monochrome Card (only 4k used)
B8000 – BCFFF	Color Graphics Card
C8000 – CBFFF	Hard Disk Controller
F4000 – FDFFF	System ROM (Usually ROM BASIC)
FE000 – FFFFF	ROM BIOS

9.5 Interrupts

The next requirement is to assign an interrupt channel. While the PC (and XT) have one interrupt controller with 8 channels, in the AT class of machines, a second interrupt controller is cascaded onto this, making for a total of 17 channels. If the interface requires interrupts, choose an available channel from Tables 9.3 and 9.4, which list the interrupt assignments on the respective classes of machines.

Table 9-3—Interrupts on PC Class Machines

NMI	RAM parity, I/O Ch. Chk., 8087
IRQ 0	Timer
IRQ 1	Keyboard
IRQ 2	Reserved
IRQ 3	RS-232C (COM2)
IRQ 4	RS-232C (COM1)
IRQ 5	Hard Disk
IRQ 6	Floppy Drive(s)
IRQ 7	LPT1: (Parallel Printer 1)

Table 9-4—Interrupt Assignment on AT-Class Machines

NMI	Parity Errors
IRQ 0	System Timer
IRQ 1	Keyboard
IRQ 2	Cascaded into the Second Controller
IRQ 3	COM2
IRQ 4	COM1
IRQ 5	LPT2
IRQ 6	Floppy Disk Controller
IRQ 7	LPT1
IRQ 8	Real-Time Clock
IRQ9	Redirected to IRQ 2
IRQ10	Reserved
IRQ 11	Reserved
IRQ 12	Reserved
IRQ 13	Coprocessor
IRQ 14	Hard Disk Controller
IRQ 15	Reserved

The interrupts marked as reserved are normally available for use by other devices such as add-on cards, network controllers, etc. IRQ 2 was marked as reserved for the EGA, but there are very few (if any) graphics cards in use that use IRQ 2. Obviously, if a card with an 8-bit connector is being used, only interrupts below IRQ 7 will be available. With this in mind, IRQ 9 is redirected into IRQ 2 in AT-class systems. In case of a clash, the user must either reassign channels or alternatively disable the interrupts of some devices. For example, the user chooses IRQ 3 and has a second

RS-232C port in the machine. Then COM2 (which uses IRQ 3) must be disabled. Otherwise, the presence of two drivers for the same interrupt may even damage the driver ICs. The device driver, when installed, will automatically redirect IRQ 3 to the appropriate handling routine, and the user need not worry about this. It is often possible to also use IRQ 7 for the interface card, since very few printer drivers actually use this interrupt, which is normally masked until enabled by an external command; this is because no interrupt channel was assigned to LPT2 in PC-class machines. It should be noted that disabling interrupts force the interface to work in the polled mode, which can considerably slow down the interface. It may be kept in mind that DOS does not support IRQ 3 and 4 for communications. Either the language in use must include driver for the COM ports (e.g., BASIC Interpreter) that supports interrupts, or the user must provide his or her own driver. Otherwise, the COM ports can be run only in a polled mode, which may not work reliably at rates above 1200 baud on PC-class machines. It may be noted that unlike ISA Interrupt sharing is possible in the MCA bus.

9.6 Direct Memory Access

The PC uses the 8237 controller, which has four channels. The assignment of other direct memory access (DMA) channels is given in Table 9.5.

Table 9-5—Direct Memory Access Channels on PC/XT

Ch. 0	DRAM Refresh
Ch. 1	Normally Free
Ch. 2	Floppy Adapter
Ch. 3	Hard Disk Controller

It is evident that while channels 1 and 3 are available on a PC, only channel 1 was available on the PC/XT. In the AT a second controller was added in a cascade configuration, providing for a total of seven DMA channels. Here, the cascading is reversed from what is done in the case of interrupts. The lower four channels are as in the XT, with an additional 3 channels on the 16-bit extension to the ISA bus. All seven channels can now address the full 24-bit address space of the AT.

In case of a clash that cannot be resolved by reassigning a DMA channel, one of the boards must have its DMA access disabled. This usually requires the removal of a jumper on the card (a minor modification) or an additional instruction to the software handler, or both. Many interface cards use DMA only in certain modes of operation and not otherwise. In such cases, once the DMA jumper is removed and as long as the user does not invoke the DMA operations, there is no change to be made. On the other hand, some drivers for cards use DMA, if available all the time. Here, in addition to removing the jumper, the software driver must also be appropriately modified. Disabling DMA can slow down block transfers considerably. On the other hand, where the transfers involve only a few bytes at a time, the speed loss may be negligible. The use of a DMA controller for data transfers involves the following stages:

1. Select the mode of operation.
2. Set the base address.
3. Set the byte count.
4. Select the memory page.
5. Enable the DMA transfer.

Actual programming of the DMA registers is fairly involved and is beyond the scope of this book. In case the interface cards support DMA transfers, the software driver or handler will include the necessary code. Interested users may refer to Chapter 10 of Eggbrecht, or Chapter 6 of Rosch, both of which have excellent discussions of DMA operations (see Suggested Reading). One interesting fact is that in an AT the DMA controller operates at half the speed of the bus, as a consequence of which, in an 8-MHz system the fastest DMA transfer will be slower than in the 4.77-MHz PC or XT. However, EISA and MCA are capable of 16- as well as 32-bit transfers. It should also be noted that DOS is limited to 8-bit DMA transfers.

9.7 Timers

Timing and counting are often required in interfacing. The PC has three channels on the Intel 8253-5. The various modes of the 8253 are summarized in Table 9.6.

Table 9-6—Operating Modes of the 8253

Mode		Functioning
0	Interrupt on Count	Counts down from the loaded preset value.
1	Hardware Retriggered Monostable	Output goes low on trigger and returns to high after preset count.
2	Clock generator	Divides clock frequency by the preset count.
3	Square Wave Generator	Produces square wave with period selected by the preset count.
4	Software Retriggered Strobe	Counts down to preset, then produces a pulse. Retriggered by software.
5	Hardware Retriggered Strobe	As mode 4 except that hardware retriggers timer.

In the PC, the clock used for the timer is 1.19318 MHz. The operation in the AT is similar. The channel assignments are as follows:

1. Channel 0 is a general-purpose timer that is set up by the BIOS. It generates IRQ 0 every 54.936 ms, or 18.206 times a second (obtained by dividing the clock by 2^{16} or 64 k). This is used by the time of day clock and floppy motor on/off controller. This interrupt is available at Interrupt Level 1C. Normally, it just returns one back to BIOS but, by installing an appropriate vector, it can be used.

2. Channel 1 is used to generate DRAM refresh through DMA Channel 0 every 15.12 μs and should not be used for any other purpose. In AT class machines the refresh frequency may be different, but the net effect is the same. Failure to refresh the RAM will have catastrophic consequences, since failure to refresh dynamic memory means the loss of the data in it. Any use of this channel must, therefore, be done with extreme caution.

3. Channel 2 is used to drive the speaker. In the original PC this was also used for the audio cassette recorder port. The output is gated under software control through bit 0 at port 0061H.

For data acquisition applications that do not use DMA, it may become necessary to switch off the time of day clock for high data transfer rates (typically 2000 bytes/sec on a 4.77 MHz XT); otherwise, the time of the day clock interrupt may interfere with data acquisition to an unacceptable degree. Applications that require fast data transfers along with a valid clock must, therefore, be run using DMA.

One point needs to be kept in mind: the DMA controller address lines can control only the lower 16 bits of the address in XT-class machines. The upper four bits are handled by a separate register called the page register. The effect of this is to restrict the DMA buffer to 64 kB, all of which must be in the same page; i.e., the DMA buffer should have the same four high bits for its address range. Using languages that limit the user to only 64 kB, it is possible to define the DMA buffer outside the area of memory normally used by appropriately programming the DMA controller. Even though DMA controllers in AT-class machines can handle all 24 lines (and EISA and 32-bit MCA can handle 32 bits), DOS restricts operation to the lower 16 bits only. The user will normally be responsible for transferring the data to his or her programs for any analysis. In BASICA, it is possible to define the segment through the DEF SEG statement and handle data acquired outside the active page with some ease. One easy way to ensure that the data area always lies in the same page for DMA transfers is to define the array to be at least twice the size required. This way, at least half of the array will always be on the same page, and this will be sufficient.

9.8 Installing the Software

After the hardware is installed the software may also have to be installed. Software for interface cards comes in two forms: handlers and divers. A handler is a program that is included in the user software and does all the accessing of the interface card resources in simple high-language commands. The driver also does the same thing in a different way. The driver gets installed as a part of the operating system at the time of booting of DOS. For this, the file CONFIG.SYS must include a statement of the form

DEVICE = xxxxx.xxx

where xxxxx.xxx is the device driver. For example, the handler for the National Instruments GPIB-PC cards is called GPIB.COM™. Then, at the time of running of the user software, a small program called the language interface must be included with the user program, which handles the transfers to and from the device driver. Starting with DOS 5.0, the INSTALL statement can be used to install run time modules into DOS itself. Installing modules permits them to be integrated into DOS (under the COMMAND shell).

In the case of handlers, the program transfers control to the handler, which in turn addresses the card. With drivers there is one additional stage, in that the language interface addresses the card through the driver. The device driver approach allows the use of the resources of DOS to the programmer, making transfers more efficient. Thus, unless some DOS features are being used, a handler will be faster than the driver due to the additional stage involved in access by the latter. However, as the device driver is common to all languages, it forces similarity in the behavior of the software interface with different languages. At the same time, the language interfaces become much smaller since they now act as interfaces between the user software and the handler. These advantages have resulted in the almost universal use of device drivers in preference to handlers. Furthermore, as the driver has to be written only once, the code can be optimized, thus, further minimizing the speed disadvantage.

Most device drivers come with their installation programs, which interactively modify the driver for the exact configuration used in that particular installation. In many cases, a utility examines the system resources and then suggests or even sets up the board. This is necessary since the exact addresses, DMA channel used, interrupt used, etc., may differ from installation to installation. With drivers, if any changes are made, it becomes necessary to reboot the system in order to install the modified driver. With handlers, either they too have editing programs to configure the exact configuration or they may need some additional routines to be used if the default configuration is changed. Since the handler is called or included only when the user software is run, it is generally not necessary to reboot the system every time the configuration is altered. Many drivers and handlers come with accompanying test programs. These should always be used, where available, to test the installation; otherwise, a series of simple programs whose outcomes are known should be run to check the installation. It may be noted that, with device drivers, some device names used in the driver become reserved (like the DOS device names). For example, if the name PLOTTER is used to define a device with the GPIB-PC driver, the system will not accept a file named PLOTTER in the same way as a file cannot be named LPT1.

9.9 Language Interfaces

When interface cards are used with drivers supplied by the card manufacturers, the programs are not all that different from those that may normally be written. On the other hand, when the cards are being directly addressed through the IO ports, the programming takes on a very different texture. The functions required are usually very poorly documented in the language manuals. The examples given are often trivial or irrelevant. For example, in BASIC one may require functions such as BLOAD, BSAVE, CLEAR, CALL, DEF SEG, DEF USR, POKE, PEEK, WAIT, OUT, INP, USR, VARPTR. Turbo Pascal (Borland) makes extensive use of PORT and MSDOS calls. Similarly, Borland C uses the *outportb* and *inportb* instructions. The best way to understand their functioning is to try out these instructions.

9.10 Working Methodology

After having selected the addresses, DMA channels, etc., one should be able to plug in the card and run. Everything must be powered down prior to inserting or removing a card. Some users are lax in this regard, but once they have blown a board or two they learn very fast. While in ISA systems, plugging (or unplugging) a board with the machine powered up may not necessarily result in shorting of tracks, on EISA boards this is almost certain to happen, with disastrous results for the board. The usual precautions that pertain to the handling of MOS devices must be taken. It is very strongly recommended that the user double check everything before installing and powering up a card. It is best if this is done by two different people. After actually installing the card, the user should install the software and run the tests. If everything is fine, he or she is in business. If not, then:

1. Make sure that all addresses, etc., are as they should be without any overlaps. One very common source of trouble is trying to map two devices into the same I/O addresses. Sometimes this may not be obvious, especially when one or more of the devices uses partial decoding and all the addresses are not given. The use

of diagnostic software such as Quarterdeck® Manifest™, Microsoft Diagnostics (MSD)™, QAplus™, etc. is invaluable in tracking these down.

2. Make sure that the correct version of DOS is being used. If software drivers are to be installed, one must have DOS 2.0 or higher.

3. Make sure that the correct procedures per the manual have been followed. Ensure that the device driver or handler is properly installed.

4. If things still do not work, seek expert help. DEBUG may be useful in identifying the problem area by decoding 00000 to 003FF. Also, many system interrogation utilities, such as Norton® SI™, Quarterdeck Manifest, QAPlus, etc., can be used to identify possible conflicts.

Suggested Reading

Eggbrecht, L. C. 1983. *Interfacing to the IBM Personal Computer.* Indianapolis: Howard W. Sams.

Krutz, Ronald L. 1988. *Interfacing Techniques in Digital Design with Emphasis on Microprocessors.* New York, NY: John Wiley & Sons.

Norton, Peter. 1985. *Programmers Guide to the IBM PC.* Washington D.C.: Microsoft Press.

Rosch, Winn L. 1992. *The Winn L. Rosch Hardware Bible 2nd Ed.* Englewood Cliffs, NJ: Prentice Hall.

Willen, D. C. and Krantz, J. I. 1983. *8088 Assembler Language Programming: The IBM PC.* Indianapolis: Howard W. Sams.

Williams, Dave. 1990. *The Programmers Technical Reference: MS-DOS, IBM PC and Compatibles.* New Delhi, India/London, UK: Galgotia/Sigma Press.

Also refer to the manuals that are appropriate to your computer.

10
The Automated Laboratory

Sanjay Gupta

10.1 Introduction

The laboratory of every generation of researchers is significantly different from that of the previous generation. Today, computers are rapidly taking over from string and sealing wax; the buzz word of the day is automation. This chapter discusses some strategies for the development of a good research and development facility with optimal investment and utilization of existing equipment.

The development of a laboratory is an evolutionary process. One cannot expect the user to learn many new tricks in a very short time. Since the operational constraints and requirements of each laboratory are unique, it is not possible to give a universal menu. The attempt here is to give suggestions, based upon extensive personal experience, that should help the user in automating his or her laboratory.

The user interface plays an important role in the success or failure of a system. Any instrumentation system, however good, is not going to be popular if the user finds it difficult to use. On the other hand, a moderate system endowed with an excellent user interface is far more likely to succeed. In modern computer-based instrumentation, this user interface is more often than not implemented through software. Thus, the development of user friendly and inherently fool-proof software becomes an integral part of laboratory automation. Some suggestions for the development of user friendly software are discussed later in the chapter.

10.2 Typical Requirements

The use of computers in laboratory automation can take three distinct forms. The strategy for interfacing will depend upon the mode of operation that is used.

In the first mode, the computer is used as a data logger, as is illustrated in Fig. 10.1. The computer simply collects data from the instrument through an interface. The computer obviously handles all data in a digital form. Analog signals may be digitized inside the instrument through IEEE 488 or digital outputs or by the computer through an analog interface card. This is often done by utilizing the recorder output of the instrument. Utilizing the recorder output of the instrument to feed the computer through a suitable analog-to-digital converter is a very common technique to conveniently get data into the computer for subsequent analysis. In this mode of operation, the data comes into the computer directly without involving the user in the painstaking and error-prone process of reading the numbers off the instrument or chart and typing them in. Once this digitized data is available in the computer, subsequent analysis is obviously fairly straightforward.

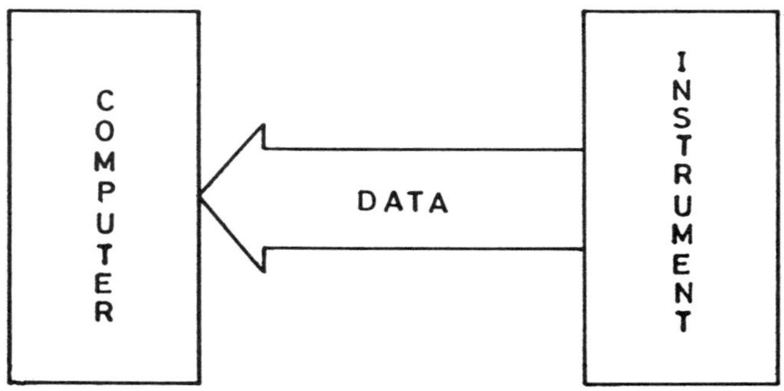

Figure 10.1—Computer as a Data Logger

The second mode of computer usage is in open-loop control (Fig. 10.2). Here, the computer performs a predetermined sequence of operations. In its purest form the computer will not try to monitor the process. This mode of operation is quite rare these days. The computer operates like a sequential programmer, i.e., sends a sequence of instructions in a predetermined sequence to initiate and/or interrupt a series of processes.

The third mode of operation is the fully integrated system or closed-loop control (Fig. 10.3), wherein the computer becomes an integral part of the control process. In this mode the actions of the control functions are based upon interpretation and analysis of the data coming in through the data logging system. This may be called the fully automated mode. As users become more and more adept at laboratory automation, their applications progressively gravitate into this third category. It should be noted that this mode of operation is not a simple combination of the previous modes. The analytical capabilities of the computer are used in a decision-making process to operate the controls.

The partitioning into the three classes is not very rigid. There can be situations in which the computer is a part of the experimental loop, but it may still be preferable to consider the situation as of the first type. For example, the computer may be used to trigger or initiate a process when a certain event occurs; thereafter, the process is

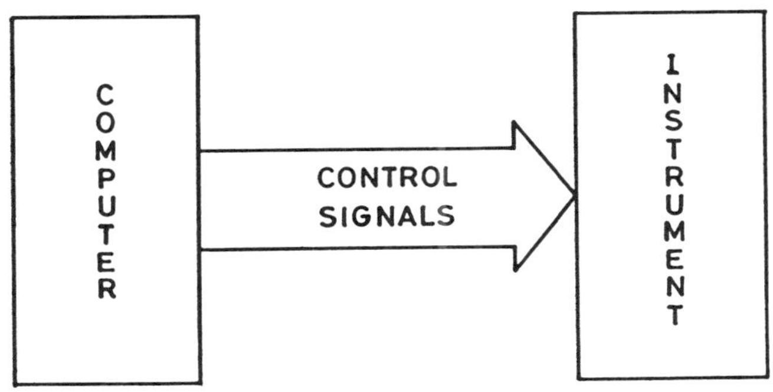

Figure 10.2—Computer as an Open-Loop Controller

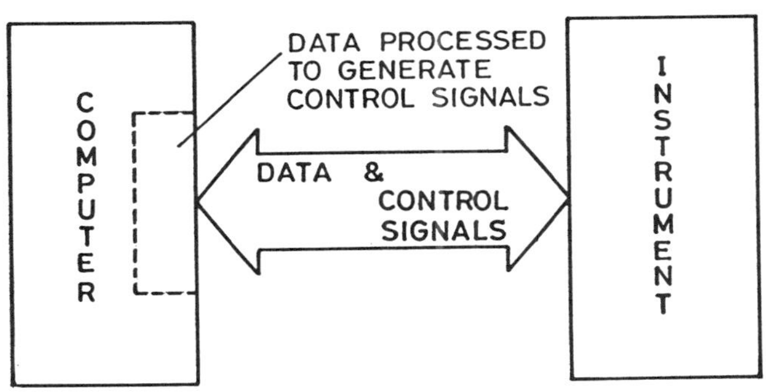

Figure 10.3—Computer as a Closed-Loop Controller

monitored with no further control action by the computer. Even though the computer is performing a control function in initiating the process, from the point of view of designing the system it will be better to treat this as a simple data logging function. This type of operation is very common in data logging from older instruments. Also, many modern, fully automated instruments can be hooked to the computer when the function of the computer is primarily to initiate the measurement and store the results for subsequent analysis. Obviously, the timing and other constraints on the system are far more severe in the last mode than in the pervious two. As a result, the computer programs also tend to be far more complex for such applications.

Depending on the complexity of the application and the interface, the applications may be categorized as follows:

1. *Data analysis.* Here, there are no interfacing requirements.

2. *Operation of single systems.* The requirements are, more often than not, simple data logging, with possibly occasional transfer of commands to the system. The interface speed is seldom of much importance.

3. *Operation of systems consisting of several smaller modules.* The computer actively participates in the control of the ensemble. The maximum benefit in instrument upgrading is often derived by automating the operation and interconnection of the various modules. The interface speed may be great importance here.

10.2.1 Data Analysis

Data analysis has been the primary use of computers in research and development from the very beginning. The use of computers for data acquisition and control is a much more recent development. Even today, the number of workers who use computers for data analysis exceeds those who use them for other purposes. The very rich and varied software for the PC makes it extremely versatile in the laboratory. It can be used for data acquisition, analysis, graphical presentation, simulation, etc. The variety of software for personal computers for graphical analysis and the display of data is unmatched.

Most experienced users always keep a spreadsheet handy for such work. Both Lotus 123 and Quattro are quite popular in this regard. The extensive library of logarithmic and trigonometric functions makes the analysis of tabular data from laboratory experiments quite easy. The extensive statistical library can be used to rapidly reduce the volume of data to manageable proportions. The available mathematical functions are more than adequate for most applications. The data can then be graphically displayed through the graphics facility of the package.

Custom packages that provide extremely powerful capabilities for data analysis are available. These range from various data acquisition and analysis packages (DaDisp™, LabView™, ASYST™, etc.) to high-end mathematical/numerical analysis packages (Mathematica™, Macsyma™, GPSS™, etc.). These are excellent packages with very powerful analysis and display capabilities. The relative rarity of similar packages does mean that one may not be able to exchange programs with other workers. On the other hand, the superb capabilities and the ease with which complex procedures can be developed and added to the library make them very desirable. CASE tools (such as National's LabWindows™) provide the flexibility of control-specific packages with the convenience of generating standard code in a popular high-level language such as C or FORTRAN. However, their cost is often high, and more common and less expensive spreadsheets can also be used for routine data analysis with excellent results.

While the use of spreadsheets is widespread and very useful for rough and ready work in situations when the same type of analysis is required frequently, it may be desirable to develop custom packages for the purpose. The development of routines for numerical analysis and graphics can be very time consuming. Fortunately, many excellent libraries of these functions are available. Users who intend to distribute software should, however, check the licensing restrictions on these packages. These can range from quite liberal, involving no royalties for executable files, to very high charges for each application being distributed. In the area of PC software, the use of DOS extenders is also often required. Due to the segment-offset method of addressing memory in the PC, quite often it is not possible to address large arrays or routines directly. Using a DOS extender on 386 (or even 286) machines allows the user to work around this problem, and, by using a combination of both a DOS extender and a virtual memory manager on a 386-class machine, it is possible to develop very large applications (4 GB if disk space permits).

For developers of graphics applications, the choice of the graphical package or library requires some thought. The package must not only be able to produce good, high-quality graphs on-screen with reasonable convenience, it must also be able to support a variety of hard copy devices. Quite often the user will require hard copy outputs on a variety of dot matrix, laser, and ink jet printers and may also want to store it for use in text-processing applications; thus, the graphics library selected should ideally be able to support the wide selection of hard copy devices with minimum fuss. Screen dumps, i.e., hard copies of screens as they appear, are not satisfactory in the modern context, since even an 800 × 600 SVGA image, when reproduced at 300 dpi, is only 68 mm × 51 mm in size! A normal A4-sized reproduction will obviously look very chunky. Thus, for good publication- or presentation-quality work, the graphics package should be able to work internally at a far higher resolution than the screen in use.

Along with the software, the user will also require display and hard copy devices for both text and graphics. The options, in order of increasing cost (and quality), are dot matrix printers, ink jet printers and laser printers. The use of plotters is declining fast as inkjet and laser printers are becoming more versatile. Plotters are being progressively restricted to exact scale drawings and the like. An inexpensive dot matrix printer is often the workhorse in the laboratory. For color output, the ink jet is often the only viable option. Color dot matrix printers are slow, suffer from poor quality, have expensive ribbons, and cannot produce overhead transparencies. Color laser printers tend to be very expensive to buy and run, and the wax transfer process used in all but the most expensive color lasers tends to make the images somewhat susceptible to damage. Quite often, users rely on an inexpensive dot matrix printer for daily work and have a laser for publication-quality work.

Another very important requirement in the laboratory is report preparation. At one time specialized packages, such as ChiWriter™, T^3™, etc., ruled the roost with their symbol-handling capabilities. The main disadvantage of custom word processing software for scientific applications is that the secretarial staff is unfamiliar with them. Currently, top-end, general-purpose packages (WordPerfect, Word, Ami Pro, etc.) also provide these capabilities. The only scientific package that enjoys widespread popularity is T_EX™, which permits device-independent formatting and ASCII files (permitting easy porting through E-mail systems), etc. However, more and more users are forsaking the ultimate flexibility of T_EX for the convenience of commercial word processing software.

10.2.2 Data Acquisition with Single Instrument Systems

Data acquisition (with or without control) with systems that consist of single instruments is quite common. These instruments may be fully functional on their own, with the computer serving only as a data storage medium and to emulate the front-panel controls for the convenience of the user. Even in this case, the capabilities may be further enhanced by using software to enhance the capabilities of the instrument. This is often achieved by on-line processing of the data and using it for instrument control. These instruments usually connect to the computer through the RS-232C or GPIB ports. As the computer works primarily as an external storage and retrieval device, the speed of the interface is not very important, and often RS-232C is the preferred interface.

Some instruments use the computer as an integral part of the instrument. In these, almost invariably a custom interface card plugs into the computer bus. In most such systems, the availability of the computer to the user for other purposes is minimal, and the computer may be regarded as a part of the system. Some instruments of this type provide the ability to connect an external computer through either RS-232C or GPIB; then, the situation is exactly as in the earlier case.

While in the case of instruments that use custom interfaces the supplier's software is obviously a must, even in instruments that use a standard interface it is usually beneficial to get the user software with the instrument, specially if the source code is provided. Such software has already been tested out and is easy to work with. It also provides invaluable guidance, when the user he wishes to develop his or her own programs.

10.2.3 Data Acquisition with Systems That Consist of Individual Smaller Instruments

Such setups are very popular in the area of electrical measurements. A test system is set up using many simpler instruments. For example, a DC resistivity measurement system will include a power supply and two meters to measure the current and voltage, respectively. Systems with four, five, or even more instruments are quite common. In these cases the use of GPIB is almost mandatory for two reasons. First, each RS-232C port can handle only one instrument; thus, one cannot handle many subsystems without additional ports. Second, the data transfer rates may be too high for RS-232C to handle. Also, in cases in which transient phenomenon are being studied, it may be desirable or even necessary to trigger a large number of subsystems simultaneously. This is possible only through GPIB. Alternatively, using non-GPIB instruments and an ADC/DAC card, it may be possible to make the required measurements. However, the use of GPIB systems is more widespread. GPIB also offers the advantage of easily mixing subsystems from different vendors.

10.3 Acquiring Instruments: Fully Automated or Modular?

Any prospective purchaser of a system is faced with a choice between purchasing a fully integrated system or building one up from individual simpler instruments. Laboratory instruments can be divided into the following three major categories:

1. Fully integrated, stand-alone instruments

2. Simple instruments, with some operations built in, that can be strung together through an external controller (the computer) into more powerful operations

3. Instrumentation systems made up of individual modules

There is considerable overlap between the three types, and each type has its own advantages and disadvantages.

Fully integrated, stand-alone instruments are usually the most expensive. The role of the external computer is often solely as a data storage medium. These instruments need not be the most powerful but are the most convenient to use. The high cost comes from the need for the manufacturer to recover the entire developmental cost from the limited number of units that are sold. Due to intense competition, the model changes are very frequent, leading to short production runs. This type of instrument is best suited for routine quality control types of work, in which similar measurements, which are well standardized, have to be made repeatedly. On the other hand, this simplicity makes it difficult to use the system in a way not envisioned by the manufacturer.

Simple instruments with a facility for full control through an external computer offer the best compromise for the research laboratory. Since the basic instrument is simple, its cost is low. With proper design of the instrument and the computer interface, it is often possible to do everything that a fully integrated instrument can do. If the instrument is not to be used very frequently, the computer can be used for other work as well. This makes it more economical. As the level of integration at the hardware level is lower, such instruments offer a much higher level of flexibility in adapting the system to other usages.

Many medium-sized instruments such as spectrometers and photometers are available in both of these varieties. The ultimate capabilities are similar in both cases, and the selection should be made on the basis of the planned end use. While the second approach is cheaper and more flexible, a dedicated quality assurance application on a production line may make the former more desirable because they are more rugged.

Developing an instrument out of individual modules is obviously a more versatile but time consuming approach. The cost may not necessarily be the lowest, since often a standard combination of functions is available for less cost than the individual functions on their own. For example, for AC impedance measurements it is cheaper to buy an impedance analyzer than to try to implement the same functions with individual frequency generators and current, voltage, and phase detectors, etc. On the other hand, a user who rarely makes AC impedance measurements, but has the other modules accessible, may connect them together and make measurements as and when the need arises.

The choice of approach should be governed by the needs; it is not possible to state that one approach is always much better than another. In general, building up a system with modules with a high level of integration may offer an economical and versatile solution in most cases.

10.4 Automation of Existing Instruments

The upgradation and automation of a laboratory is an evolutionary process, and there are bound to be instruments of varying vintages in any existing laboratory. Some of these will not be bus-based. The potential for their interfacing may differ widely. It is by no means necessary to consign these instruments to the junk heap, since they can be used for good purpose. For these older instruments, an analog and digital interface card (ADC/DAC) is necessary. Once the computer is equipped with digital and analog interfacing capabilities, it is possible to extend the useful life as well the capabilities of the older generation of instruments.

Most medium-sized instruments (spectrometers, thermal analyzers, photometers, etc.) have recorder outputs. This type of system is ideally suited for connecting to computers used as data loggers. It is usually a simple task to feed these recorder outputs into the analog inputs of the interface card. Once this is done, the data can be acquired and analyzed using the computer. In this case, the equipment can have a performance that is comparable to the equivalent instrument of the fully integrated type.

Many smaller instruments (pH meters, electrometers, etc.) also have analog outputs for recorders. These outputs should be used for feeding the ADC of the interface card. Some instruments also have digital outputs (very often BCD). If the levels are TTL or can be converted to TTL, these can be used to drive the port in the input mode on the interface card. However, in many cases the protocol requirements can be rather difficult to handle, and it may be easier to use the analog output instead.

Digital displays are very popular in many simple laboratory instruments. Many of these instruments use a single chip converter and display driver. 7106 and 7107 are quite commonly used ICs for this purpose in 3-1/2 digit displays. These ICs can directly drive a display with a maximum voltage of ±200.0 mV. Thus, the analog part of the instrument is designed to convert the final value in the 200 mV range, which is then fed to the display driver. On the 7106, pins 30 and 31 correspond to the negative and positive inputs, respectively. Other ICs will use different input pins, but the principle remains the same. This signal can often be tapped for computer interfacing with minimal or no changes in the electronic circuit of the instrument. Care must be taken not to overload the electronics and also to protect the instrument against signals from the leads. This is easily done by putting suitable resistors (and a capacitor) on the leads. For signals less than ±600 mV, two silicon diodes in an antiparallel configuration provide additional protection (see Fig. 10.4). It is also a good practice to use the diodes to protect the computer against high voltage signals when thermocouple signals are being fed in.

Many power supplies, temperature controllers, etc., have a remote programming capability. This may be either through an external potentiometer or voltage. A little study of the circuit will reveal that in both cases the actual programming is through a voltage. The DAC output(s) of the interface card can be used for this. In some instruments this facility is apparently not there. However, the setting knob on the front panel may be operating in this mode. It is then a simple matter to substitute the setting signal with the DAC output.

In every case the user must ensure that the voltages and currents that are present will not damage the instrument or the computer. The user will be well advised to consult a person who can confirm that no damage will result from the planned connections. The author has used these methods with great success without damaging any instrument or computer.

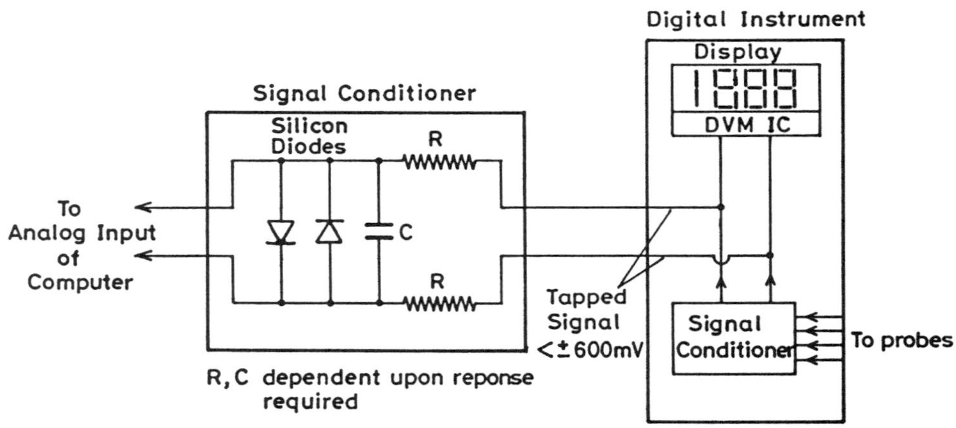

Figure 10.4—Tapping of Low Level Signals

10.5 Choice of Interface: RS-232C or GPIB?

Many instruments offer the option of either using RS-232C or GPIB. Quite often there is a straight option, or one or the other may be available as an extra cost option. If the computer in the laboratory has a GPIB card in it and the cable length limitations (2 m per instrument, 20 m maximum) are not a hindrance, then GPIB should be preferred. Remember that while one RS-232C port can handle only one device, a single GPIB port can handle up to sixteen devices; thus, GPIB is much less costly in terms of resource utilization. Recently, in process control instrumentation the use of RS-422 (an updated variant of RS-232) is gaining ground, since it permits multiple instruments to share the same lines, i.e., a multidrop configuration. The data transfer rates under GPIB can be as high as 3 MB/sec (300 kB/sec typical), while under RS-232C the rates are limited to 9600 baud, with many software packages limited to only 2400 baud. These rates correspond to under 1 kB/sec.

Another problem with RS-232C on the PC lies in the driver. DOS does not include interrupt support for the COM ports; i.e., although interrupts are assigned for COM1 (IRQ 4) and COM2 (IRQ 3), these are not utilized by DOS. Only the BASIC interpreter includes a driver for the COM ports that supports the interrupts. In later versions of DOS, COM3 and COM4 are also supported, but they share the interrupts with COM1 and COM2, respectively. Users of other languages must, in general, include their own drivers in order to support these interrupts. The use of these ports using the DOS driver (i.e., in the polled mode) at faster than 2400 bps is not advisable.

Therefore, on the basis of convenience and transfer speed, GPIB wins hands down. However, RS-232C does offer certain advantages. It is cheap; every computer has one or two RS-232C ports (and up to four are possible), and the cable can be an inexpensive twisted pair in many cases. Most stand-alone instruments are connected on their own to the PC with at most a printer (usually on Centronics) and a mouse (on the other RS-232C) and, if the computer is going to be dedicated to that instrument, the extra cost of GPIB cannot be justified. Also, RS-232C, unlike the GPIB, does not have any serious distance limitations. However, on the whole, if available, GPIB is usually the better option.

10.6 The User Interface

Any software system is ultimately going to be used by humans. The ergonomics of the package go a long way in deciding its acceptability to the user. A well-defined system will not only perform its task well but will also be one that is convenient for the user to work with. The user interface includes both the data acquisition program and the subsequent reporting. The techniques of report preparation are well known to most experienced programmers and will not be discussed here. The problems of the user interface for data acquisition and control are outlined in this section.

The user interface needs special care for any process-related application. In the area of process control it is almost certain that the software will be used at some point in time by persons who are not experts and may not even be very conversant with computers. Sometimes the user may have been working continuously for long periods and may be very tired. At other times things may not be going smoothly and the user may be irritated or upset. Under such conditions the chances of human error are increased manifold. Also, as the software is used over and over again, any minor inconvenience will grow in the user's mind. Therefore, great care must be exercised in designing the human interface. Experience has shown that the amount of program code required to implement the user interface often exceeds the functional part of the program. The interaction of the user with the system can be divided into three distinct stages:

1. *At start-up*: At the beginning the user is required to provide the various parameters for the program to function. This is done before the execution of the main body of the program starts.

2. *On-line display*: During the actual execution of the program some parameters have to be displayed. Alarms or warnings for unusual operating conditions may have to be given.

3. *On-line interaction*: It may become necessary to take inputs or request some manual actions from the user during the execution of the program. Furthermore, the user may need to interrupt the process or change some parameters.

The computer screen should be kept uncluttered as far as possible. The programs must be able to tolerate wrong key strokes, erroneous data, etc., without causing disasters. The level of error proofing required in a program depends upon the frequency of use, skills of the operators, etc. As operator skills go down, the requirements on the user interface become progressively more stringent. However, poor programming practice is certain to land the user into trouble at some time or another. Some principles that will help the user avoid most of the pitfalls are presented below.

10.6.1 At Start-up

At the beginning of any application the user is normally required to define the initial operating parameters for the program. If only one or two parameters are to be sought, the problem is simple. Quite often a large number of parameters are required to be read in. In such cases a hierarchial menu structure is necessary. Some basic guidelines for the user interface prior to the actual operative part of the package are listed below:

1. A *sign-on* message giving the name of the package, its function, names of the developers, version, and date should be provided.

2. Parameters that characterize the process as a whole and are rarely changed are best provided to the program through a data file. These process constants differ for different systems and, hence, cannot be made program constants. For example, in a program to control furnaces, the process constants for a particular furnace once determined remain invariant, but each furnace is different. The furnace characteristics are best stored in files, and the data file is input through the menu.

3. The menu structure should be hierarchial. Parameters related to a single function only should appear on any particular screen. The name of the parameter should be unambiguously displayed. If possible, the default value and/or the acceptable range should also be indicated.

4. The data entered should be validated. If an invalid input is received, the user should be asked to enter the value again. An error message may also be displayed, if appropriate.

5. A pathway should be provided at each level of the menu to return to the previous level. Furthermore, often it is desirable to validate the screen before proceeding to the next screen. In case the number of parameters is very large, it may be desirable to validate the entire set before proceeding to the execution of the process.

Most developers develop their own set of routines for routine operations such as file opening, data entry, data validation, etc., which they then incorporate into various programs. This not only avoids repetitive development of code but also provides consistency across the programs of the particular developer. The C code of one such library is given in Fig. 10.5 below. The routines handle data input within specified limits with or without allowing default inputs, validation, buffer clearing, etc. Such interfaces become almost essential if one is to avoid run time errors due to incorrect inputs, etc.

```
double get_real(double min,double max, int zero);
int get_integer(int min,int max, int zero);
void draw_box(int x1, int y1, int x2, int y2,int opmode);
char validate(void);
void clear_buffer(void);

double get_real(double min,double max, int zero)
/*   read in a 'double' number between (min<value<max)
     if zero!=0 then allows a <CR> to return a 0 value, otherwise not */

{
    double value;
    char a[80];
    char far *errptr;
    int x,y,a_len;
    x=wherex(); y=wherey();
    do
    {
        gotoxy(x,y);clreol();
        gets(a);
        a_len = strlen(a);
        if((a_len == 0) && (zero != 0)) {value = 0; return(value);};
        value = strtod(&a[0], &errptr);
    }
```

```
    while((errptr != &a[a_len]) || (value < min) || (value > max));
    return(value);
}

int get_integer(int min,int max, int zero)
/*   read in an integer between (min<value<max)
     if zero!=0 then allows a <CR> to return a 0 value, otherwise not */
{
    double value;
    int x,y,a_len, int_value;
    char a[80];
    char far *errptr;

    x = wherex(); y = wherey();
    do
    {
        gotoxy(x,y);clreol();
        gets(a);
        a_len = strlen(a);
        if((a_len == 0) && (zero != 0)) {int_value = 0; return(int_value);};
        value = strtod(&a[0],&errptr);
        int_value = value;
    }
    while((errptr!=&a[a_len]) || (int_value<min) || (int_value>max));
    return(int_value);
}

void draw_box(int x1, int y1, int x2, int y2,int opmode)
/*   Draws a box in character window from (x1,y1) to (x2,y2)
     uses double lines if opmode=0 otherwise uses single lines */

{
    int i;
    char a[20];
    char far *str1 = "╔═╗║╚╝";
    char far *str2 = "┌─┐│└┘";
    if (opmode==0) strcpy(a, str1); else strcpy(a, str2);
    gotoxy(x1,y1);
    printf("%c",a[0]);
    for (i=x1+1; i<x2; i++) printf("%c",a[1]);
    printf("%c",a[2]);
    if ((y2-y1)>1)
    {
```

```
        for(i=y1+1; i<y2; i++)
        {
            gotoxy(x1,i);printf("%c",a[3]);
            gotoxy(x2,i);printf("%c",a[3]);
        }
    }
    gotoxy(x1,y2);printf("%c",a[4]);
    for (i=x1+1; i<x2; i++) printf("%c",a[1]);
    gotoxy(x2,y2);printf("%c",a[5]);
}

char validate()
{
    int x,y;
    char ch;
    x = wherex();
    y = wherey();
    do
    {
        gotoxy(x,y); clreol();
        ch = getch();
        ch = toupper(ch);
    }
    while ((ch != 'N') && (ch != 'Y'));
    if (ch == 'N') printf("No"); else printf("Yes");
    return(ch);
}

void clear_buffer()
// clear the keyboard buffer
{
    while( kbhit() ) getch();
}
```

Figure 10.5—C Library for Some Common Functions

10.6.2 On-line Display

Once the various parameters have been input, the actual execution of the control program begins. Here, the important thing is to display the process information in an appropriate form. In general, the display must show the current state of the system and the parameters of interest. It should be kept in mind that if the program is executing over a long period of time, the operator not only needs information about what is going on but also an assurance that everything is fine. Also, the operator cannot be expected to concentrate on the display for hours on end. There are many ways to design a user friendly screen. Some of the important considerations are as follows:

1. Display only those parameters that are of immediate interest. If required, more parameters can be made available to the operator when required. This may occur through the use of command keys or may be displayed automatically in a window as and when required. For example, if unusual operating conditions are detected, a window displaying the relevant parameters may automatically open up.

2. Make creative use of screen attributes to indicate the status. Highlighting, inverse video, blinking, and color (if available) can be used to indicate abnormal operation, request user intervention, etc.

3. The sound generation capabilities should be used to attract the attention of the user. This may be done to indicate faulty operation of the process as well as special periods of activity. In addition to the standard beep <CTRL>G (ASCII 07), the sound generator can be used to produce a wide variety of sounds. However, unless all operators are musically inclined, more than two or three frequencies should not be used.

4. Various operator intervention options should be displayed.

5. The use of graphic displays may be desirable in some cases, provided sufficient computing capacity is available. While graphic displays look nice and indicate trends well, they are not very useful for providing the exact value of the process parameter.

10.6.3 On-line Interaction

During the execution of the process, provision for intervention by the operator must exist. This intervention may be required to interrupt or terminate the process, to change some parameters while the process is going on, or to input information to the program during execution at predefined stages. Some important aspects in the design of the interaction routines are as follows:

1. If the process is not to be interrupted during the user intervention, writing a keyboard parser becomes necessary. This routine stores the keystrokes in an

array, and after the operator signals the end of the intervention, it interprets the array and changes the parameters. The normal sequence of the intervention is identifier, value, terminator. The identifier indicates the parameter that is to be modified. Next, the value is entered; finally, the terminator (often <RETURN>) indicates the end of the entry. Once the terminator is received, the parser interprets the value and takes appropriate action.

2. If the process is to be interrupted during the user intervention, then before proceeding to the data entry menu, it must be put into a safe state; i.e., the control outputs set so as to prevent the possibility of disaster if the operator does not respond to the prompts in time.

3. Validation should be requested, especially if the operator action will have disastrous consequences for the process at hand; e.g., if he or she wishes to end the run.

4. At the end of the run, the system must be put into a predefined safe state.

5. Single keystroke interruptions should be avoided. These should never be used for critical operations. It is possible that one key may be pressed by mistake, leading to fatal consequences for the process. One simple way is to use the function keys in combination with the <SHIFT>, <ALT>, or <CONTROL> keys. Programs that stop as soon as any key (unspecified) is pressed can be even more dangerous.

6. If necessary, help menus should be provided. Following the lead of Lotus 123, "Hit <F1> for Help" has become almost universal. Recognizing the need for *on-line help,* many compilers provide help compilers that make this task simple.

7. The user interface should be consistent across the entire menu structure. Using Windows (or other GUIs), this is almost automatic, but many developers may not wish to get involved with the complexities of GUI programming. However, if object-oriented methods are used, a certain degree of consistency across the various levels of menus becomes automatic.

There can be many user interfaces for the same application, all of which may be equally good. Each user tends to develop a basic structure of the user interface, which is used in almost all of his or her programs. It is often convenient to divide the screen into sections for various functions. For example, the top left corner maybe reserved for a program identification message. Often, the time of day is displayed in the top right-hand corner. The bottom two or three lines may be reserved for user interaction. The rest of the screen is then used for the display of parameters, which may be alphanumeric, graphic, or both.

Suggested Reading

Barney, George C. 1985. *Intelligent Instrumentation: Microprocessor Applications in Measurement and Control*. London, UK: Prentice Hall Intl.

Hunter, Ronald P. 1987. *Automated Process Control Systems, Concepts and Hardware*. Englewood Cliffs, NJ: Prentice Hall.

Krutz, Ronald L. 1988. *Interfacing Techniques in Digital Design with Emphasis on Microprocessors*. New York, NY: John Wiley & Sons.

Keithley, J. F., Yeager, R. R., Erdman, R. J. 1984. *Low Level Measurements*. Cleveland: Keithley Instruments, Inc.

Morris, Alan S. 1993. *Principles of Measurement and Instrumentation*. London, UK: Prentice Hall.

Zaks, Rodney. 1980. *Microprocessors*. Berkeley: Sybex.

11 Real-Life Applications and Cautions

Jai P. Gupta

11.1 Introduction

Computers can, in principle, be used to facilitate, automate, or eliminate every action that a human being takes. The practicality of it is currently restricted due to the limitations of computers on memory, speed, lack of thought processes, voice recognition, expression of emotions, reading from text or drawings, and the like. However, artificial intelligence (AI) experts are working on all these fronts and are very hopeful of success during the next few years in some of these areas and, during the next few decades, in many others.

Before the use of computers, the control strategies were by and large limited to ON/OFF, P, PI, and PID. The hardware costs made it prohibitive to implement the more sophisticated strategies unless absolutely necessary. With the computing power now available it is possible to use more advanced techniques, such as feedforward and adaptive control, reliably and economically. The development of fuzzy logic controllers provides faster and smoother control in many applications that range from high speed trains to complex industrial processes.

Turning to the microcomputers that form the basis of this book, their capabilities are being continuously expanded by developments in the hardware (processors, peripherals, interfaces, etc.) and in software (languages, tools, methods, etc.). An example can be cited: The computing capabilities of a mainframe computer of the early 1960s were far less than those of a digital diary today. The capital and operating costs (manpower, space, repair and maintenance, air conditioning, training, etc.) have all come down very significantly, making the unit affordable and operable by almost everyone who wishes to do so [1].

With the availability of such a powerful tool, where and how to use it is left to the individual's imagination. In a few years, the question of its "applications" would become as silly as currently asking the applications of paper (writing, printing, currency notes, threatening letters, paper aeroplanes in class rooms) or of the plastics. Be that as it may in the future, in this chapter we will discuss some applications in brief so as to meaningfully synthesize the separate discussions in the earlier chapters. These are related to scientific and industrial (engineering) applications. The all-important applications in inventory control, finance, accounting, human relations, defense, etc., are not considered here because of the specialized nature of the book. See the literature for publications that deal with these topics.

The microcomputer applications that will be discussed here are:

1. control of a batch distillation column,
2. control of a bioreactor (fermenter),
3. control of a furnace, and
4. monitoring of a long pipeline.

As already stated, these few applications are only the proverbial tip of the iceberg. The applications are, in fact, unlimited, right from monitoring the fetus through all phases of human active and retired life to prescribing an up-to-the-last-minute medical treatment plan.

The wide applications noted above naturally bring into question the reliability of the computer system, i.e., the hardware, the software, and the liveware (designers of the systems and users/operators). It is obvious that the computers will control more and more complex situations as time progresses and the cost of such control systems continues to decrease. Already, the designers of such systems are generally not the end users, and the final system is almost invariably assembled by integrating hardware and software units as well as devices from different vendors. Most users/operators of such systems believe that the systems are infallible without having a real basis for such a conclusion and without knowing what is inside the black box. Such a situation is fraught with immense dangers. Hence, one should exercise appropriate caution in using computerized controls. The last section of this chapter deals with these aspects.

11.2 Control of a Batch Distillation Column

Distillation is the process of separation of a multicomponent liquid mixture into pure constituents. The actual purity of the products is dependent upon several factors that include the nature of the components, the purity desired, and the economics of the process among others. Distillation can be continuous or batch-wise. Continuous distillation is used when the demand for the product is high, as in petroleum refineries, petrochemical complexes, etc. These achieve a sort of steady-state operation, wherein a mixture (or crude oil) is continuously fed in and the products are continuously withdrawn from the distillation column. This makes for a relatively easier operation and, hence, easier control.

Batch distillation is used when the demand for the product is limited, as in specialized chemicals and dyes, the pharmaceutical industry, etc. These do not achieve steady state because the mixture is fed in only at the beginning and the purified product is withdrawn continuously, thereby changing the composition of the remaining mixture in the column at any given time and reducing its amount as the distillation continues. It is evident that the batch distillation column is more difficult to control since the aim is to obtain as consistent a product purity as is possible with changing composition and temperature in the column. The control scheme for this should be considered at the column design stage, because it will influence the design of the column and the control components. Any modifications and/or alterations in the control scheme or in the column can be made more easily at the design stage than after the column has been fabricated.

A batch distillation column is shown in Fig. 11.1. A reboiler at the bottom supplies the heat of vaporization to boil off the mixture in the reboiler. Initially, when the mixture contains significant amounts of the more volatile (lower boiling point) components, their concentration in the vapor phase predominates, and, consequently, the temperature of the vapor leaving the reboiler is also low. A condenser atop the column removes the latent heat from the vapor leaving the top of the column and condenses it. A part of this condensate is returned to the column as a reflux, while the balance is withdrawn as the desired product. The liquid reflux entering at the top flows down the column and contacts the vapor rising from the reboiler at the bottom. This contact is made on all the trays in the column, and, depending upon the nature of the components, their relative amounts, and the temperature at each of the trays, an exchange of components occurs between the two phases. This makes the rising vapor richer in the more volatile components and the down-flowing liquid richer in the less volatile ones, thus helping achieve better separation and a higher purity product.

As the mixture becomes depleted of the more volatile components, the boiling temperature of the remaining mixture in the column rises, as does the temperature in the reboiler. The temperature profile in the column keeps changing, as does the composition of the vapor reaching the top and becoming condensed in the condenser. This results in a continuous change in the product composition and, hence, in its purity. It initially contains more of the more volatile components, and gradually, as their amounts in the column deplete, the product composition shifts towards the less volatile (higher boiling point) components. This can be corrected by continuously increasing the amount of the reflux instead of keeping it constant. This is a difficult proposition at best and would result in a decreasing amount of product being formed with the passage of time. The practical way out is to change the reflux ratio at definite times (after a preliminary analysis of the system) so that the final product mix obtained at the completion of the batch distillation is of the desired purity.

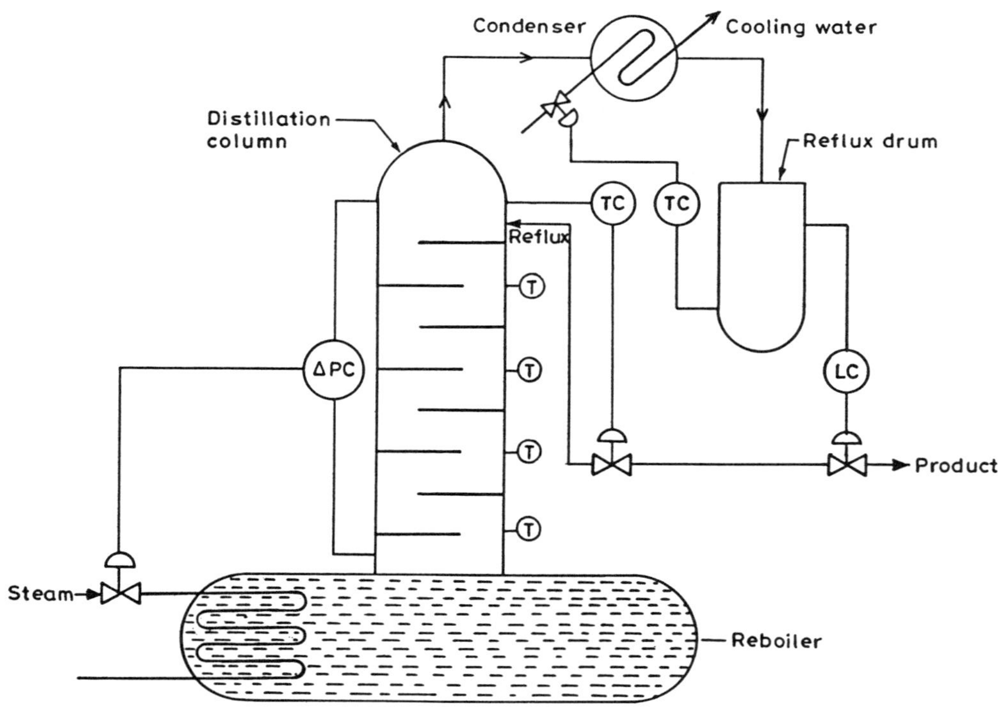

Figure 11.1—Simplified Control Diagram for Batch Distillation Column

It is evident from the above discussion that one needs to control the following in batch distillation:

1. The heat input to the reboiler to control the rate of vaporization so that there is sufficient time for the vapor and liquid to come to equilibrium at each of the trays for effective component separation. The pressure drop in vapor flow through the column is small, which otherwise would adversely affect the working of the column. This heat input can be controlled by adjusting the pressure of the steam in the reboiler heating coils or the current to heaters if it is an electrically heated reboiler.

2. The heat taken out at the condenser with a change in the overhead vapor composition. This will also affect the reflux temperature and, hence, the separation achieved at each of the trays. This can be done by controlling the flow rate of the

cooling water rather than its temperature, which is a more involved process for large quantities.

3. The reflux ratio, i.e., the ratio of the amount of reflux to that of product withdrawn from the condenser. This affects the purity as well as the amount of the product. This can be done by adjusting a valve in the reflux return line that goes from the overhead condenser to the distillation column.

Some of the control elements are also shown in Fig. 11.1, which is a somewhat simplified representation of the real situation. Pressure drop across the tower is related to the vapor flow rate and, hence, is used to control the steam pressure to the reboiler. (The pressure drop in the condensate reflux is relatively low since its volumetric flow rate is very low as compared to that of the vapor. Besides, it will not matter since the pressure taps are kept in the vapor zone.) The temperature of the condensate is used as a measure to control the cooling water flow. The temperature at the top of the column is a measure of the composition of the vapor coming out and is used to control the reflux ratio. The use of sensors to measure the temperature at each (or select) trays gives a temperature profile in the column that can be related to the composition of the mixture and, hence, to the effective separation achieved.

Since the signals generated at the sensors and those needed to activate the final control elements are analog, several analog-to-digital and digital-to-analog converters (ADCs, DACs) are needed. The pneumatic control valves for the steam to the reboiler heating coil, cooling water supply to the condenser, and at the reflux ratio control require electropneumatic converters. All these are standard items and are available commercially.

In this case as well as in all other cases used in the laboratory or industry, one would require appropriate software. This will have a control algorithm for the specific situation, taking into account the time delays and time constants for the process; A-to-D conversion of a signal and its equivalent values in engineering units, taking into account any nonlinearity in the relationship between the signal and the actual value as in the case of thermocouples used to measure temperatures; and D-to-A conversion of the computer response and activation of the control elements and/or alarms as the case may be. Sometimes the signal obtained from the sensors may be filtered to remove any noise so that the control action is smooth. For further details, References 2 through 4 may be consulted.

11.3 Control of a Bioreactor

Bioreactors, also generally known as fermenters, are used in the production of pharmaceuticals, drugs, ethanol, and specialized products. With the current hi-tech involvement in genetics, the importance of bioreactors has increased significantly, as has the importance of their more precise data collection, analysis, and control.

A bioreactor has several parameters to control [5]. Some, such as the temperature, pressure, agitator power, foam level, turbidity, etc., affect the physical environment, while others, such as the pH, dissolved oxygen, dissolved carbon dioxide, sugar level, protein level, etc., affect the chemical environment. To these may be added the control of some specific ions (Na^+, Mg^{++}, K^+, PO_4^{3-}), DNA, ATP, and so on, for more precise chemical environment control. The specific sensors for measuring the various parameters are discussed in Reference 5.

The limitations (as they exist in developing more precise data logging, analysis, and control) relate not to the computer hardware but to the availability of accurate, stable, and fast-response sensors of different types and to the software to take into account the numerous specific interactions and interdependences of the various parameters for each fermentation process. Much effort is being exerted in these areas with remarkable success, with more to follow. Eventually, a computer should be able to optimize the various parameters and control strategies to evolve a better fermentation process.

The discussion here will be restricted to the control of pH, temperature, and dissolved oxygen DO as developed in the author's laboratory. The bioreactor setup is shown in Fig. 11.2 [6]. The system studied was the fermentation of molasses to produce Baker's yeast. The setup consists of a 14-liter bioreactor with pH and dissolved oxygen probes and a sheathed thermocouple dipping into the fermentation broth. The pH and the temperature were controlled in an ON-OFF manner, using a solenoid valve for the pH and turning on the circulation of refrigerated water for the temperature, while the dissolved oxygen was controlled by proportional-integral-derivative (PID), proportional-integral (PI), and proportional-only (P) control strategies, using a stepper motor-driven valve. This procedure was adopted so that a variety of control formulations were available, as the need arose. These strategies can be easily used in any other chemical engineering control setup.

In the fermentation of molasses, pyruvic acid is evolved, which lowers the pH while the heat released during fermentation increases the temperature of the broth. Oxygen is consumed during the process of cell growth. Hence, the pH was maintained by using dilute aqueous ammonia, the temperature was maintained by circulating cold water, and the dissolved oxygen was maintained by bubbling sterilized air into the broth.

A simplified flowsheet of the software is shown in Fig. 11.3 [7]. One can choose any of the P, PI, and PID control strategies for the dissolved oxygen. The set value of any of the controlled variables (pH, temperature, and DO) as well as the dead band (or

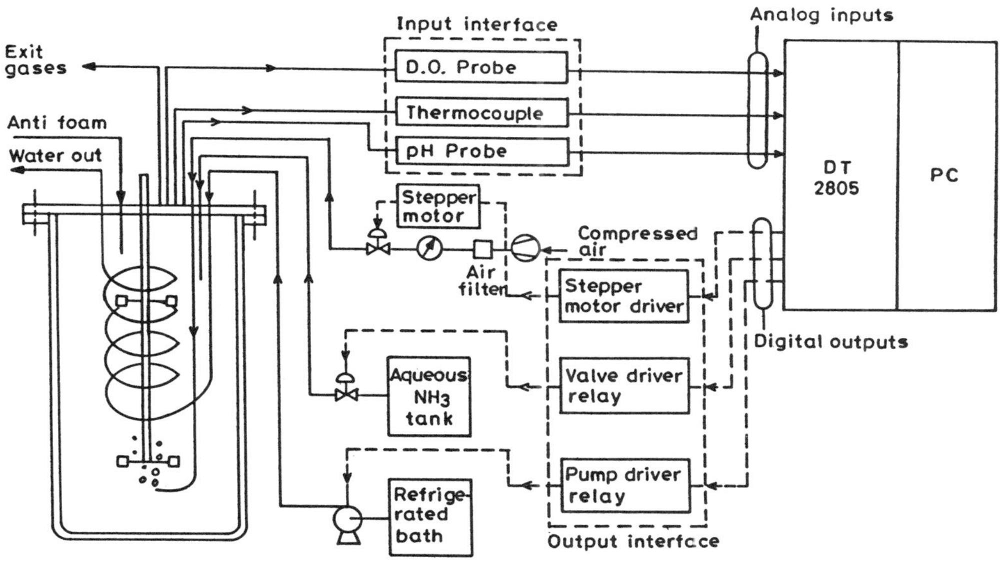

Figure 11.2—Bioreactor Setup [6]

margin of permissible error) can be changed any time during the run. The currently sensed values and the position of the valves (ON/OFF for pH and temperature, and % OPEN for dissolved oxygen) are displayed on the screen and printed on the dot matrix printer. These are also stored on a floppy for later use if need arises. The history of the controlled parameters since start-up can be plotted on the screen and printed on the plotter as a screen dump (see Fig. 11.4). More parameters for control and interaction can be built-in to form a dynamic control strategy. Audiovisual alarms can also be incorporated. These may be of more use in a fast chemical reaction system than in a slow fermentation one. The total cost of a dedicated microprocessor-based control system is very moderate.

11.4 Control of a Furnace

Furnaces are used extensively in the chemical and metallurgical industries as well as in laboratories in a variety of situations (heating of a material or fluid stream, boiling of water and other liquids, cracking of naphtha, heat treatment of equipment for stress relieving, melting of ore, etc.). These may be heated electrically, or by gas or fuel oil. Temperatures range from a couple of hundred degrees Celsius to 1200°C

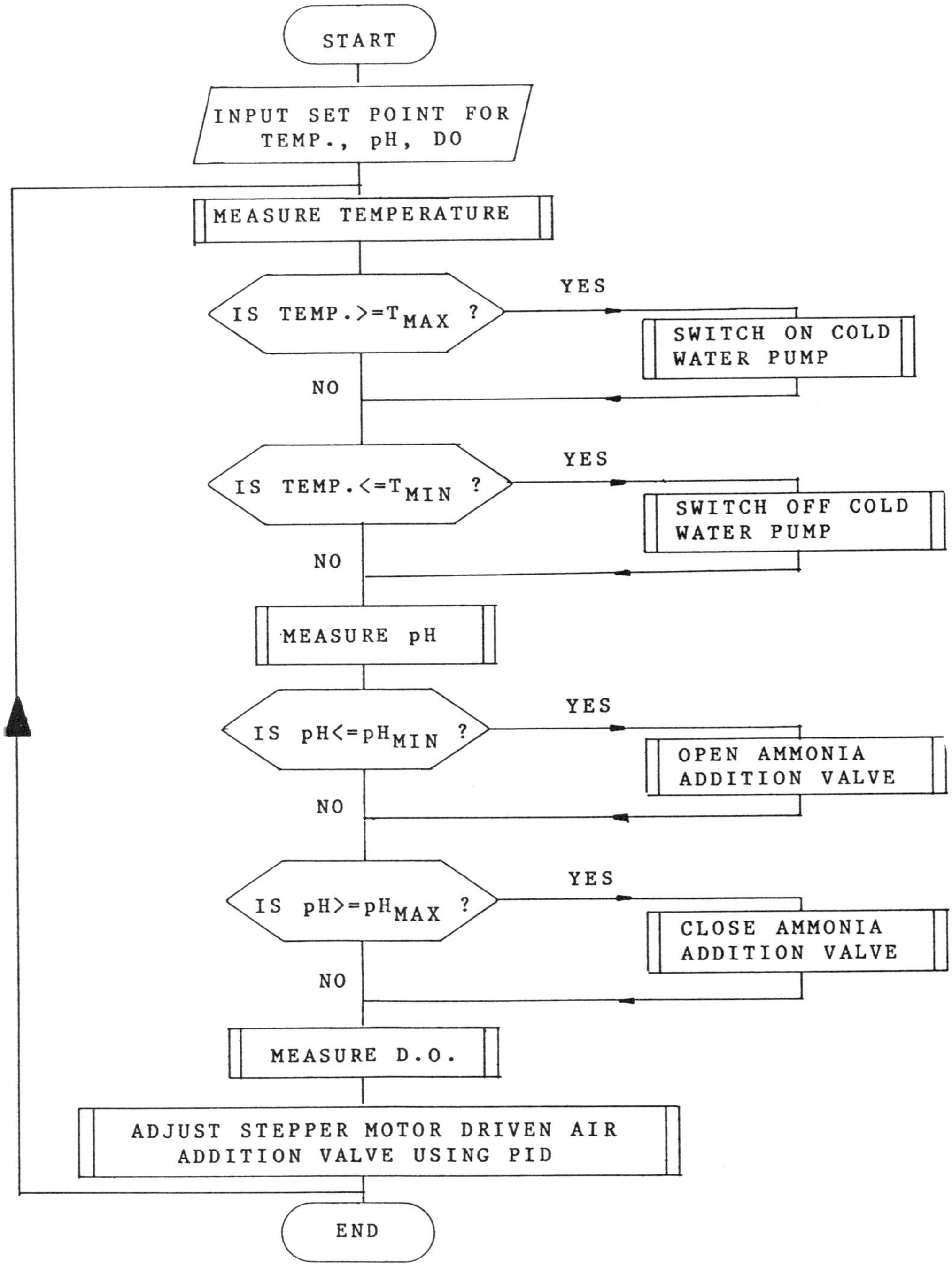

Figure 11.3—Simplified Flow Sheet of Software for Bioreactor Control [7]

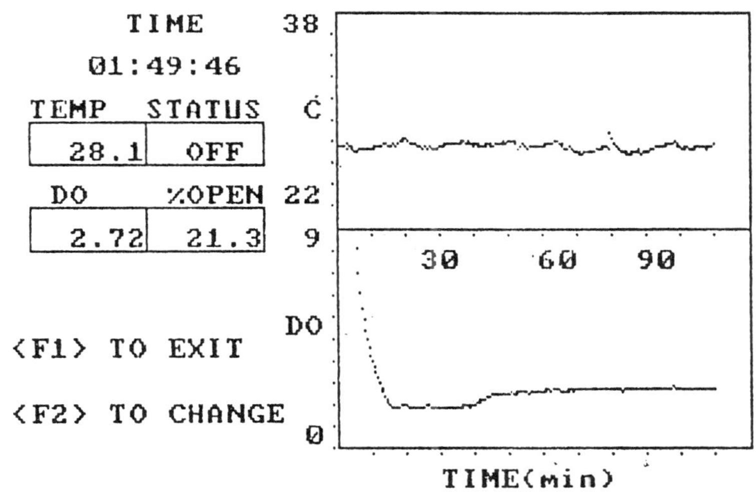

Figure 11.4—Bioreactor Control Display

and higher. It is obvious that the control of furnace temperature is a very important parameter for the furnace to do its job efficiently and optimally. Concerns for energy conservation have also put increased emphasis on good temperature control to save on power and fuel.

The discussion below is for the control of a laboratory tubular furnace heated electrically [8]. The temperature is sensed by a single thermocouple placed strategically in the furnace. The algorithm developed can be expanded for a larger furnace in which the temperature is sensed at several locations and pooling of information is done at the control point. Individual heaters (or gas/oil burners) need to be controlled in different zones to obtain a uniform temperature in such a case.

The software discussed here is capable of controlling the temperature of a furnace to within ±1°C. It can also be programmed for heating and cooling rates (within the limits of 100% and 0% power inputs) to an accuracy of ±3°C at a heating rate of 120°C/hr. The difference is due to the time lag of the system and is well within acceptable limits for most operations. It will reduce with the lowering of the heating rate. The software can be programmed to heat a furnace in a step-wise fashion, i.e., specific heating rate to a point, holding there for a fixed time, heating again at a given rate (may be different from the first leg) to another temperature, and holding there for a fixed time (may be different than the first holding time), and so on (see Fig. 11.5). Such a situation is often encountered in metallurgical operations, the study of alloys, phase transition, crystal behavior, and so on.

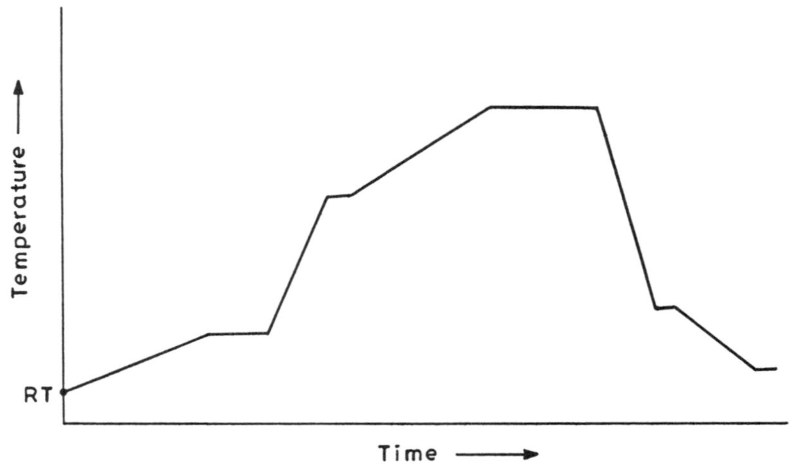

Figure 11.5—Temperature vs. Time Trace for Furnace

Simple on-off control is not good because of the large lag time in any furnace. Lag time extending to several minutes can damage the material being processed in the furnace, since the temperature overshoot can be disastrous. Large oscillations in temperature can result. If a proportional-only type of control is used, the offset is significant and the controller response rather sluggish. A PID algorithm gives a rapid response with no offset and minimum oscillations or hunting. Hence, the software developed has a PID algorithm with a loop gain and time constant for the integral and derivative controls. These parameters can be obtained by the process reaction curve method of Cohen and Coon [9].

The heating and cooling kinetics are different since the cooling is usually passive, depending upon the surrounding atmospheric conditions, while heating is less affected by the surroundings. However, in practice the parameters estimated for heating can also be used for cooling a furnace without significant errors, especially for a laboratory-size furnace. A sketch of the setup is shown in Fig. 11.6, and a short flowsheet of the computer program in Fig. 11.7. A single PC-based system, as developed here, can control up to seven furnaces simultaneously. The restriction to seven is due to the DT2805 card used, which has a total of 8 different A-to-D converter input channels. One of these is dedicated to the room temperature sensor, which is used for cold junction compensation.

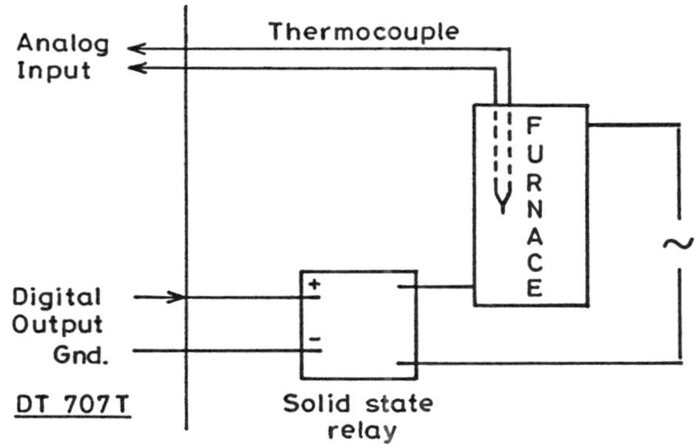

Figure 11.6—Connection of Furnace to Interface

The output of the computer using the PID equation has to drive some control element in the system. One can use phase control, voltage control, and time control. The first two (i.e., phase control and voltage control) require additional hardware and are not linear in response, requiring a correction before the drive is generated or errors will be introduced. Therefore, time proportioning is preferred, since it is linear and requires only an inexpensive solid-state relay as an interface between the furnace and the computer-generated control signal. One should choose the time step or interval judiciously. If it is too small, the PID calculations by the control algorithm will be done very often, and the random errors in temperature measurement can lead to wrong or spurious control signal generation. On the other hand, if the time step is very large, the control will be poor. A 30-second time interval between successive calculations of the control signal was found to be optimal for typical tubular laboratory furnaces (1.5 kW, 300 mm long, 30 mm diameter). For large furnaces, this time interval may be suitably increased.

The furnace control discussed above has been implemented using an PC with Data Translation's DT2805 Low Level Analog Interface Card™ with its companion DT707T Screw Terminal Panel™. The DT2805 can be replaced by any combination of a temperature-measuring device and a single channel of digital (TTL) output suitably connected to the computer. The software is written in Turbo PASCAL Version 3.0, using a locally developed driver for the DT2805. The program control loop takes about 0.45 second (on a PC running at 4.77 MHz) and needs to be called only once every 30 seconds. The driver takes less than one ms and needs to be called once every second.

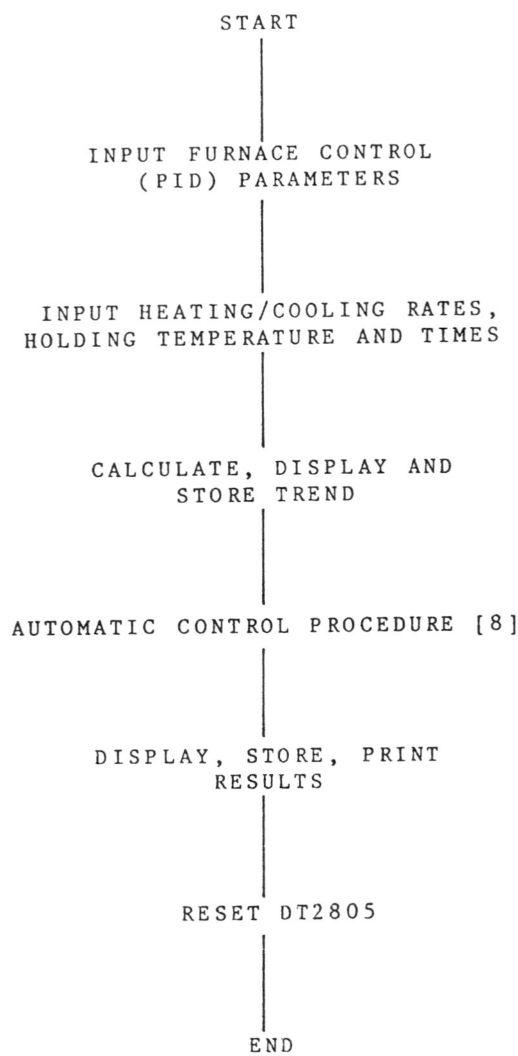

Figure 11.7—Short Flow Sheet of Furnace Control Software

11.5 Monitoring of Long Pipelines

Long pipelines are used to transport crude oil, petroleum products, natural gas, etc., within a nation, across international boundaries, or along the sea bed. These play an important role in the overall industrial scenes of many nations. Leaks in such pipelines can result in heavy financial losses due to lost raw material and products, interruptions of production, the high fire and explosion hazards that such leaked material would pose. Such hazards are also injurious to living beings. Hence, it is imperative that leaks in such pipelines, due to design errors, natural causes (corrosion, earthquakes, etc.), or acts of sabotage, be detected very early, the leaking section isolated, and predetermined corrective actions taken in as short a period as possible to stop the leak. The repair and investigation can follow thereafter.

All the fluids travelling in a pipeline are sent under pressure to overcome the friction or drag in the pipeline and any changes in elevation along its path. Leakage brings a part or the whole of the stream of this fluid into the atmosphere, the amount depending upon the size of the leak. There is a sudden drop in the pressure of the fluid. As a result, a pressure pulse or a shock wave travels up and down the pipeline from the point of leakage (see Fig. 11.8). The velocity of the pressure wave is equal to the velocity of sound in the fluid. The leak detection consists of measuring this pressure wave, discounting any spurious events, isolating the leaking section by closing the valves both up and down from the leaking point and nearest to it, sounding alarms in the control room and pumping station (or compressor houses in case of a gas), informing the plants upstream and downstream of the leak, and rerouting the flow if an alternate bypass line exists around the leak. All of these actions have to happen automatically. Even after this, the monitoring of the pipeline has to continue since it still contains the fluid, unless the whole of the fluid has been cleared out by driving a sphere through the line or by means of a compressed inert gas [4].

The software for pipeline monitoring is rather sophisticated in that the partial differential equations for mass, energy, and momentum need to be evaluated. The thermophysical properties of the fluid (i.e., density, viscosity, thermal conductivity, heat capacity, compressibility, etc.) have to be known since they vary with both pressure and temperature. The actual pressure loss between two consecutive monitoring stations and the time taken for flow should also be known so that the peaks that result due to the leaks can be easily detected. The changes in the cross-section of the pipeline, in elevation, in temperature (underground, over the river bed, directly exposed to sun and rain), and any side withdrawals or branch lines also affect the pressure values monitored.

Microcomputers are used at each monitoring site, while mainframe units are installed to monitor the whole pipeline. The microcomputers communicate with the main frame (or supervisory) computer as well as with each other through networking. In order to safeguard against a spurious trip, a voting system using three units is used so that the action is initiated only when at least two out of three microcomputers get the input that shows a leak. Separate sensors are used for each microcomputer, or the failure of the single sensor will cause a common mode failure of all the three microcomputers (and the sense of security will be a false one).

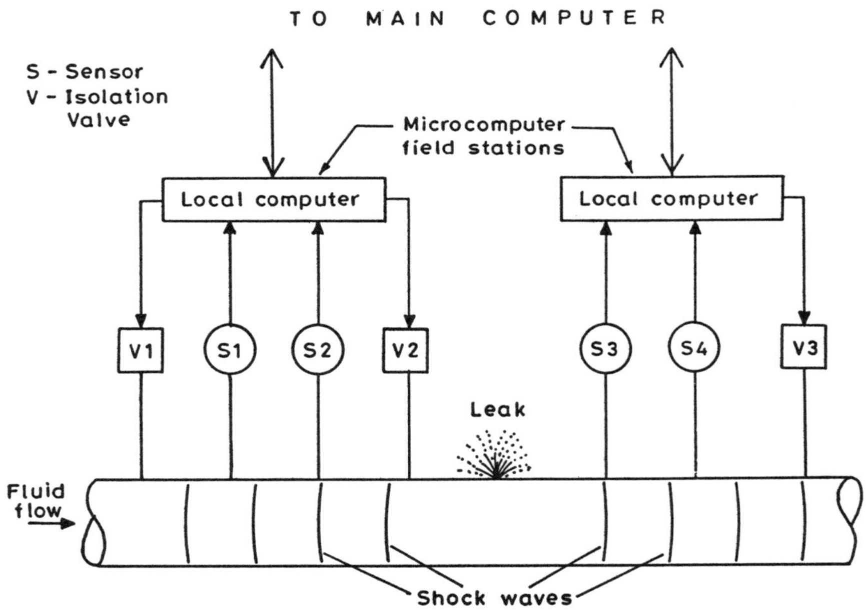

Figure 11.8—Simplified Diagram of Pipeline Control

The polling of the three microcomputers can be done locally and a decision to initiate actions communicated to the main control room for any final decision (to delay by a short time, rechecking, overriding if the leak is really small and stoppage of flow will cause greater harm, etc.). Alternately, the polling can be done at the main control room itself, where data from all other field stations would also be coming in regularly. This would avoid the probability of failure of the polling system at the site stations, while more data will need to be transferred to the main station, which may necessitate a more powerful control computer and more sophisticated (hence, more expensive) data link (buses and communication are discussed in Chapter 6). The choice is actually dependent upon where the decision makers are located if a fully automatic, stand-alone system at each site is not used. If such stand-alone systems are used, the decision to take action can be made at each site and the actions initiated. If the decision making is not done locally, the reliability of the data link to the master controller becomes even more critical. With longer and larger diameter pipelines being laid and with exciting developments in microcomputer networking and related electronics fields, the exact formulations of the monitoring and control systems will continue to get updated, more efficient, reliable, and more cost-effective with time.

Another approach to avoid spurious trips is to make more than one type of measurement. In a pipeline system, these could be the pressure pulse (shock wave) with a noninvasive flow rate sensor or two detectors for pressure pulses in close proximity. Only if both the sensors show an alarm situation would the emergency action in a preset sequence be initiated; otherwise, it would not. It is assumed that the probability of both the sensors failing together (fail safe) is almost negligible, and the planned changes in the flow rate due to side withdrawals are subtracted and do not confuse the flow rate sensor by using suitable software. (In the case of a chemical reactor, the two variables can be the temperature and the concentration of a specific component or the thermal conductivity/electrical conductivity of the fluid.)

The best approach is to use two different signals to avoid a transducer failure sending spurious message to the microcomputer and a majority voting system using three microprocessors to avoid a microprocessor failure.

The computer will determine, based on the time the signals are received and the locations of the detector stations, the section of the pipe in which the leak has developed. Then the further steps in the emergency shutdown will start as numerated in the beginning, i.e., isolating the leaking section, bypassing it if an alternate route exists, slowly closing the valves (quick closing will produce further shock waves), informing the upstream and downstream plants and users, and shutting down the total pumping if need be. (There may be situations when total pumping is not shut down if, between the feed point and the leak point, some user plants are located and the leaking section can be isolated without disturbing the supply to these plants.)

In critical situations, such as pipelines, the software will have to be such as to allow time for each action to be completed in sequence (or in parallel, as the case may be) and ensure completion of the action (i. e., not leaving any valves only partly closed, etc.). It should be kept in mind that a background noise would exist in pressure measurements due to the pumps and compressor, which force the fluid through the pipe. This should be subtracted out by using a pair of pressure transducers at a specific distance apart.

Very small continuous leaks, such as at flanged joints or pipe welds embedded in soil, which develop during the transients at start-up, may go undetected for long periods of time. If the pipe is above ground or in water, the leakage of this type would eventually be noticed.

Since pipelines are very long, data gathering can pose problems. Hence, a distributed system is required with outstations/field stations to gather data from the sensors. The data is transferred by dedicated telephone cables, a microwave link, or a satellite link. Surveillance is provided by field staff as well as by overflying helicopters.

11.6 PCs for Industrial Applications

Most industrial applications of the PC do not require the desktop version used in offices and laboratories because of several reasons, two of which are [10] as follows:

1. The PC cannot take the hostile environment of a factory floor with all its chemical vapors, dirt, fumes, vibration, higher temperature, electrical noise, physical abuse, etc.

2. The industrial user may not need the keyboard, VDU, speakers, or even the disk drives.

Units are now available (at a cost) that can take the hostile environment of a factory floor and work at 55°C or even higher without air conditioning. They can tolerate such conditions far longer than can a human being [10]. These units can be floor mounted or put into industrial panels, NEMA enclosures, or standard racks. A combination of PC/ATs are being used to control factories for which complete packages, including hardware and software, are available. Some of the functions they perform in parallel are data acquisition, command, data processing, trending, alarm management, and supervisory control [11]. Users of such systems are naturally more concerned with system capabilities, reliability, and ease of learning than with the bits and bytes or the hardware details. To reduce the space requirement and the heat generated, miniature PC cards are also available.

The developments taking place in the use of PCs in industrial control are really phenomenal by any standards. Better, more versatile, more reliable, and easier to use and cheaper cards and software are being put on the market with greater speed than one can comprehend, as a look at the several journals and magazines in the area of PCs will reveal. Up to 8 color coded "pens" per trend and two trends per display screen can show the history of up to 16 variables at any moment. Software has been developed to show animated graphics of the process, real-time trend, and longer-range historical trend of several variables all at one time on the VDU [12]. One PC can be linked to as many as 30 different recorders and data loggers via an RS-422 line. Each front-end can monitor 512 points, and, hence, one PC can monitor 4096 points. Larger systems can be configured with a local area network (LAN) [11].

Most of the better industrial controllers, which are invariably microprocessor-based, also allow a change of parameters and control techniques at the push of a button.

The software procured from the market should preferably be a standard one from a reputed supplier. Such software is well documented, maintained, and updated; customer support is provided; and, most importantly, it is well tested before it is released on the market. The testing may have been on a P&I diagram with one or a few loops at a time being tested, or a simulator may have been used to test the entire plant with all the possible control situations. The latter is to be preferred for the confidence it gives even though it is costlier. One should also find out:

1. how easy is it to configure the system using the particular software,

2. if the process can be simulated,

3. how long the training takes, and

4. how difficult is it to make changes.

Hardware reliability is also very essential in order for the control strategy to work. Even if the main computer system is reliable, the operation can be affected adversely by (1) frequency changes or spikes in the power supply that will corrupt the memory, (2) the failure of input sensors that causes loss of signals, (3) the printer getting jammed, (4) loose connections, etc. One must build in systems to check against these and provide for alternate methods or alarm sounding. On-line system diagnostics is a very helpful tool, since it can detect faults to the board level besides monitoring the internal status of the system periodically while the computer is on the control job.

11.7 Cautions in Using a Control System

Computer-aided production is attractive; computer-aided disaster is abhorrent.

Users want a control system that has a very high degree of integrity and dependability under the most exacting conditions and at an affordable cost. This requires that the control system should both operate when needed and not operate when not needed; i.e., spurious operations should be avoided and the system should be designed to fail safe and not fail to danger. It must be realized that the control systems and/or their sensors and connecting cables will need to work in the extreme conditions of temperature, pressure, dust, fumes, chemical vapors, humidity, etc. Such systems must be reliable even with transmission line noises, voltage fluctuations, and the like. Further, they should not permit deliberate or accidental settings of control parameters beyond a safety range.

Some of the following information has been taken from References 13 through 15.

The various possible faults in a control system can be classified into the following types:

1. *Hardware faults*. The equipment does not perform as expected. Such faults with computer hardware are not unknown, though a far larger number of faults occur

with the equipment, valves, sensors, and measuring devices that are attached to the computers.

2. *Software faults*. These occur due to errors in the instructions given to the computer. They can be both in the software obtained from vendors or the one developed in-house. Thorough testing is essential under all possible operating conditions and combinations thereof. For critical controls, the simulation of conditions should be as thorough as is done in flight simulators, since the stakes can be as high or higher.

3. *Treating the computer as a black box*. This tendency needs to be checked right from the first use of computers that one makes. It is essential that the results obtained be critically evaluated as to whether they make sense or not. This could happen due to faults in the software, hardware, or input data. In engineering, it is vital that engineers get into the habit of quickly estimating an approximate number around which they expect the results (similar to the way one generally does a quick estimate of the expected restaurant bill or a grocery bill before the store clerk has made one). If the value is very different than expected, one should do a more through analysis before proceeding any further.

4. *Misjudging operator response*. Since most operators have very limited knowledge about the computer controls they work with, they tend to become disinterested and may often bypass or ignore an alarm situation, especially if it occurs frequently. It is important that operator training be thorough and updated with the controls. Operators may occasionally be tested or quizzed to get an idea of their grasp and then be retrained if necessary.

It is essential that the ultimate safety protection should be based on an independent computer system and not be part of the control system computer, lest the former should also fail along with the latter.

In most cases, the errors are not of the computer hardware but of the human failure to address the situation. Operators are not told of the changes made in input, control settings, operating instructions, batch composition, input rates, etc. They work in a vacuum. Their lack of training about the awesome control systems makes them less interested and bored.

11.7.1 Computer Control in Chemical Plants

For the use of computer control in chemical plants, a basic and strict discipline is required in design, selection of items from reputed vendors, installation, thorough testing, operation, and maintenance of the controls. This is because any slackness at any stage can lead to disasters. The following are some points of concerns:

1. Inappropriate/incomplete specifications and design of the control system
2. Procurement of poor quality hardware and software
3. Unsatisfactory or incomplete validation of the software
4. Installation of such controls in a less than satisfactory manner
5. Testing and maintenance of the controls not fully documented
6. Inappropriate training or not updating of all the personnel involved in operation and maintenance

For computer control systems to work, the following step-wise process should be followed [14]:

1. Assess the hazards in the process [17].
2. Decide the necessary safety requirements to take care of the hazards
3. Decide the required level of safety (frequency and severity of the hazards permitted or tolerable)
4. Design the computer control safety system for it
5. Conduct a safety audit
6. Check to ensure that the required level of safety will be met

Reference 16 also lists the guidelines for the operation, maintenance, and modification of the computerized safety systems as well as for the personnel training to deal with them. These are very important, and users should develop the habit of following these guidelines as well as any that are recommended by the designers, vendors, inspectors, etc., because the chemical industry and its controls are becoming more and more sophisticated.

A new emerging trend is telecontrol (controlling from a distance) which could also mean from another country or across the oceans. The author has visited one such facility that maintained several thermal power plants and a nuclear one. The implications are numerous, and the lines of responsibilities and legal obligations are blurred. It is hoped that they will become clarified with time and as experience in using such services is acquired.

With advancements in computer systems, the scene is moving from quantitative methods in design and operation to value judgments via expert systems, artificial intelligence, and knowledge-based techniques. In these, the subjective judgments of the designers of the systems will play a major role in their use and, hence, in the safe operation of the plants.

In the emerging scenario, it is important that the total control regime be considered together so that the various electronic, optical, and mechanical sensors and devices (actuators, valves, controllers, etc.) can be designed to support each other, since the safety requirements of different areas of a plant may be different. Once experience is gained and analyzed, the control engineers will be ready for the next generation of controls.

Modern trends tend to intensify process disturbances due to continuous high throughputs, intense operating conditions, recycles, and the elimination of intermediate storage by linking plants to form an integrated whole. Besides, the system is required to provide a 7×24 service (7 days × 24 hours) and, hence, higher reliability is essential.

Some problems in computer control are:

1. bus latency and response time,
2. integrity of the data highway,
3. data highway bandwidth for the worst scenario, and
4. due to automatic control, the operator is often required for monitoring only, which may lead to boredom and/or apathy that can be bad in case of an alarm situation.

Perturbations can arise due to:

1. raw material quality and availability,
2. service quality and availability (steam, air, water, power, gases, etc.),
3. plant equipment availability and efficiency, and
4. environmental conditions.

11.7.2 Possible Problems and Cautions

1. Voltage fluctuation may bypass some instructions in the software.

2. Real-time analysis may not have been adequately done. The computer, after instructing a valve to close, goes on to the next step without allowing time for the mechanical valve to close and testing that it has closed.

3. Software modifications can introduce serious faults. These should be treated as seriously as the plant modifications, with all the possible consequences analyzed before the modifications are done.

4. If a manual override or parallel operation is possible, it may leave the automatic system in an unsafe position if the manual system has been used to turn the operation off. The wrong status of the operation can play havoc when the automatic system is put back on without first updating it for the results of the manual override. Any use of manual override should be properly documented, told to the supervisor and the next shift operator, and the automatic system adjusted for its effects.

5. While analyzing relative costs of controls and other options, the total lifetime costs should be kept in mind. This includes regular testing, maintenance, and repair or replacement.

6. The operators should be told of the disastrous consequences of doing anything in an unauthorized manner. This includes unauthorized testing of alarms and trips, set points, valves, and sensors or carrying out any tests, etc. The Chernobyl nuclear accident is reported to have occurred, as per one school of thought, because of unauthorized tests.

7. In HAZOP of computer control systems one should rigorously analyze (a) the consequences of errors in data entry, (b) real-time operation not being completed, (c) voltage fluctuation in the line or due to lightning, (d) the slowing of the PC due to parallel connections to other divisions of the company even if for a short time, (e) more than one emergency situation arising at the same time, (f) the presence of viruses in the software, (g) manual overrides, and so on.

11.8 Conclusions

In this chapter, four very different control applications of small computers have been discussed. This should give some idea of the variety of situations in which PCs are used. The configuration for actual use on the floor in a plant is simpler but more rugged due to the hostile working environment. The microcomputer control has to be quite reliable. This requires that both the software and the hardware be thoroughly tested under the extremes of the operating conditions.

References

[1] Gupta, Jai P. November 1986. "Microprocessor Control Applications," in Lecture Notes for *Winter School on Microprocessor Applications in the Laboratory*, pp. (XI-1)–(XI-10). Co-ordinators: Sanjay Gupta and Jai P. Gupta, Kanpur, India: I.I.T. Kanpur.

[2] Zoss, Leslie M. 1979. *Applied Instrumentation in the Process Industries*. Houston: Gulf.

[3] Buckley, Page S.; Luyben, William L.; and Shunta, Joseph P. 1985. *Design of Distillation Column Control Systems*. Research Triangle Park, NC: Instrument Society of America.

[4] Robinson, E. R. 1985. *Microcomputers in the Process Industry*. Chichester: Ellis Horwood.

[5] Aiba, S.; Humphrey, A. E.; and Millis, N. F. 1973. *Biochemical Engineering* (2nd ed.). Tokyo: University of Tokyo Press.

[6] Kakkar, S. 1987. "Microcomputer Based On-Line Control and Study of a Bioreactor." Master's Thesis. Coguides: Jai P. Gupta and Sanjay Gupta, Kanpur, India: Chemical Engineering Dept., I.I.T. Kanpur.

[7] Desai, A. 1987. "A Microcomputer Controlled Bioreactor: A Comparative Study of Control Schemes for the Control of Dissolved Oxygen in Molasses Fermentation." Master's Thesis. Coguides: Jai P. Gupta and Sanjay Gupta, Kanpur, India: Chemical Engineering Dept., I.I.T. Kanpur.

[8] Bhatnagar, S. 1988. "Computerized Acquisition and Analysis of Impedence Data: A Software System." Master's Thesis. Coguides: K. Shahi and Sanjay Gupta, Kanpur, India: Materials Science Program, I.I.T. Kanpur.

[9] Cohen, G. H., and Coon, G. A. "Theoretical Consideration of Retarded Control," Trans. ASME, 75: 827-834.

[10] Babb, M. 1986. "Riding on the PC Bus: New Avenues to Industrial Automation," *Control Engineering*, 33, 10: 64-65.

[11] Babb, M. 1987. "PC Software: Affordable Automation Solution," *ibid* 34, 10: 37-39.

[12] Babb, M. 1986. "Blue Collar Software Packages Help Process Engineering Get Back in Control," *ibid* 33, 10: 57-60.

[13] Kletz, Trevor A. 1991, "Human Problems with Computer Control: An Update," *Plant/Operations Progress,* 10, 1: 17-21.

[14] Jones, P. G. 1991. "Computers in Chemical Plants—A Need for Safety Awareness," *Trans I Chem E,* 69B: 135-138.

[15] Gupta, S. November 1992. "Control System Design for Loss Prevention in Process Industries," in Lecture Notes for Short Course on *Hazard Analysis in Chemical Industry,* pp. LP-1–LP-5. Coordinator: J.P. Gupta, Kanpur, India: I.I.T. Kanpur.

[16] "Programmable Electronic Systems in Safety Related Systems," Parts 1 and 2 *Health & Safety Executive*, 1988. London: Her Majesty's Stationery Office.

[17] Gupta, Jai P. 1993. "Quantitative Risk Assessment and Hazard Analysis in Chemical Industry," Short Course, Kanpur, India: I.I.T. Kanpur.

12
Laboratory Exercises

Sanjay Bhatnagar and V. C. Srivastava

12.1 Introduction

This chapter gives some simple laboratory exercises with their accompanying computer programs. These can be used as a guideline for developing exercises using the facilities that may be available in any particular laboratory. The computer programs given have all been tested. In order to illustrate the use of more than one language, all programs using the National Instruments GPIB-PC card are in Turbo PASCAL and those using the Data Translation ADC/DAC/DIO card are in BASICA.

It is recommended that the laboratory exercises follow a cyclic approach. Here, each problem is broken up into stages of increasing complexity, with each level building upon the previous level. This approach ensures that the student gets a grasp of at least all the basics, even if he is unable to complete the entire problem. A three-stage approach to a temperature control problem may proceed as follows. In stage one the student is asked to make a temperature measurement and to separately switch a heater on, then off. In stage two he may be asked to combine these into a simple on/off controller. In stage three the control may be enhanced to include a dead band, with a proper display of process time, temperature, and status. If needed, more sophisticated control schemes may be introduced at this stage. This way anyone who completes even part one has an idea of analog input and digital output. An additonal advantage of this approach is that the problem itself becomes the instruction sheet for working it out.

One golden rule for developing any interfacing software is to declare and initialize all parameters that may be required by the interface routines. BASIC and FORTRAN programmers must be very careful in this regard and guard against typographical errors. PASCAL programmers are automatically protected against typographical errors but must be careful about the initialization. These elementary precautions can save many hours of wasted effort.

The following exercises are illustrative of simple uses of interfacing. Before each exercise the hardware requirements are given. The user may develop variants depending upon the hardware available with him. In each case the complete program

---

Readers interested in obtaining programs given in this chapter on floppy disk may contact the editors.

has been given for stage 3. In case of difficulties in stages 1 or 2, the program listing for stage 3 should be consulted for guidance. It is assumed that the manuals for the various cards and instruments used are handy.

12.2 Simple Operation of HP 7470A Plotter

Hardware required:

- GPIB-PC Interface Card with Turbo PASCAL Interface
- HP7470A Graphics Plotter (HPIB version) installed as PLOTTER.

This problem is designed to teach simple operations of the HP7470A Graphics Plotter under GPIB. This device supports the Hewlett Packard Graphics Language (HPGL). Thus, almost any plotter supporting HPGL can be substituted for the 7470A. A simple exercise in three stages is given below. In this exercise only, the commands required in stages 1 and 2 are also given.

Stage 1: Using IBIC send commands to do the following:

1. Find the plotter [IBFIND PLOTTER]
2. Select pen 1 [IBWRT "SP1;"]
3. Move the pen to a point (250,350) [IBWRT"PU 250,350;"]
4. Lower the pen and draw a line by moving the pen to (2500,2500) [IBWRT "PD 2500, 2500;PU;"]
5. Replace the pen in the stall [IBWRT "SP0;"]

Stage 2: Using IBIC as above do the following:

1. Draw a box from (300,400) to (3000,4000) [IBWRT "SP1;PU300,400;PD300, 4000,3000,4000,3000,400,300,400;PU;SP0;"]
2. Write your name starting at (1500,500) [IBWRT "SP1;PU1500,500, LBmyname \3;SP0;"]

Stage 3: Write a computer program to:

1. Define a plotting window of (0,0) to (10300,7650)

2. Write X-AXIS along the X-axis
3. Write Y-AXIS along the Y-axis.
4. Write EXPT.1 in large letters on the top.
5. Draw a box for the plotting area of (400,600) to (10000,7000)
6. Write HP7470A in the eight directions of the compass.

The Turbo PASCAL program is given below:

program EXERCISE_1;

```
(*  Program to illustrate simple operations with the HP7470A Plotter
    No error checking is done
       Uses IBFIND, IBWRT from Turbo Pascal Driver
            IN, PU, PD, LB, DI, SI, SP functions of HPGL     *)

  {$I TPDECL.PAS  { Include Turbo Pascal declarations}

  type tempstr=string[255];
  var i,dev:integer;
      st1,st2:string[5];
      ter:char;

   procedure LOADSTR(var outst:iobuf;instr:tempstr;len:integer);
    (* Function    :    Takes in a string and loads it in an array  *)
      var i:integer;
      begin
        for I:=1 to len do
          outst[i]:=instr[i];
      end;

   procedure SBWRT(dev:integer;message:tempstr;len:integer);
    (* Function    :  Takes in the command as a string argument and
                      sends it to the device through IBWRT function *)
     var wrt:iobuf;
     begin
       loadstr(wrt,message,len);
       ibwrt(dev,wrt,len);
     end;

  begin
    ter:=chr(3);
    dev:=ibfind('plotter');    {Initialize plotter address}
```

```
    sbwrt(dev,'in;',3);  { INITIALIZE AND RESET}
    sbwrt(dev,'ip 0,0,10300,7650;',18);  {Define window}
    sbwrt(dev,'si .38,.54;',11);   {select character size}
                    {now label the axes and write header}
    sbwrt(dev,'sp 1,pu 4600,280,lbX-AXIS'+ter+',di 0,1;',34);
    sbwrt(dev,'pu 250,3100,lbY-AXIS'+ter+',di;',25);
                    {note 03hex as terminator}
    sbwrt(dev,'sp 2,pu 3500,6000,si 1,1.5,lbEXPT.1'+ter+';',37);
                    {now draw the frame}
    sbwrt(dev,'sp 1,pu 400,600,pd 400,7000,
              10000,7000,10000,600,400,600;',56);
    sbwrt(dev,'si .38,.54,pu 5000,3500;',23);

    for i:=0 to 7 do          {Directions of the compass}
      begin
        str(cos(i*arctan(1)):5:3,st1);
        str(sin(i*arctan(1)):5:3,st2);
        sbwrt(dev,'di ' +st1+ ',' +st2+ ';',15);
        sbwrt(dev,'lb  HP7470A'+ter+',pu 5000,3500;',25);
      end;
    sbwrt(dev,'sp 0,pa 0,0;',12);
  end.
```

The plot is shown in Fig. 12.1.

It should be noted that a procedure sbwrt has been defined, which simply loads the ASCII string passed on to it into the array, which in turn is transferred to the plotter by the GPIB-PC handler function IBWRT. This routine is more of a convenience since it is easier working with a string than with an array. The program initializes the plotter through the IN command to clear any errors, and at the end of the program parks the pen in the holder through SP0. The extensive use of strings in GPIB operations is obvious.

12.3 Use of ADC Card as Autoranging DVM

Hardware required:

- DT2805 input-output card with PCTHERM software
- Variable voltage source

This problem is designed to teach the user the following:

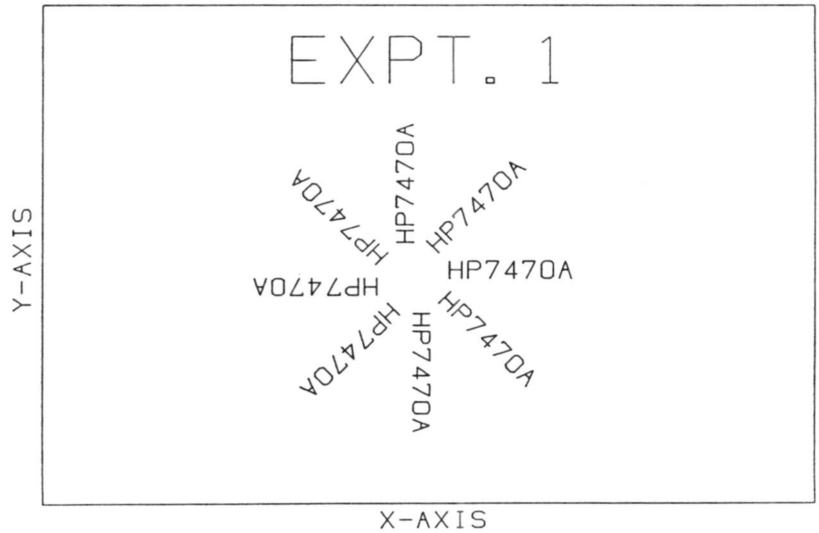

Figure 12.1—Plotted Output in Exercise 12.1

1. Analog-to-digital conversion

2. Use of the PCTHERM software interface for the DT2805 low level analog and digital IO card on the PC

3. To implement an autoranging function

The exercise in three stages is given below:

Stage 1: Write a BASIC program to:

1. Load the PCTHERM driver, using DEF SEG and BLOAD

2. Measure the voltage at input channel 4 using the XAV command. The value measured should be checked using a multimeter or the XMV command to verify the correctness of the conversion equation.

Stage 2: Implement autoranging by measuring the voltage with a gain of 1 and then use the measured value to set the gain at 1, 10, 100 or 500, as appropriate. Measure and display the voltage along with the gain used.

Stage 3: Combine the above programs to set up an autoranging voltmeter. Generate a sound of a frequency of (50 + absolute voltage in millivolts). The pitch will vary as the voltage applied is changed.

The BASICA program for this is given below. It should be noted that an ADC value of 0 corresponds to –10 V for a gain of 1. For other gains it is appropriately scaled.

```
10 '**************************************************************
20 '*                                                            *
30 '*          EXERCISE TO DEMONSTRATE AUTORANGING               *
40 '*             AND ANALOG.DATA.VALUE CONCEPT                  *
50 '*                    Name,   Date                            *
60 '*                                                            *
70 '**************************************************************
80 '
90 '  uses DT2805 with PCTHERM
100 '       SOUND used to generate a sound with a frequency
110 '              FREQ = Voltage (mV) + 50  Hz
120 DEFINT X   'To simplify use of PCTHERM
130 DEF SEG = 7552
140 PRINT "loading PCTHERM... "
150 BLOAD "pctherm",0
160 PRINT "initializing.....":GOSUB 1000
170 CALL XAV(CH%,GAIN%,ADV%) 'get ANALOG DATA VALUE
180 '
190 '*****   AUTO GAIN SETTING     *****
200   GAIN%=1
210   IF ABS(ADV%-2048)<204 THEN GAIN%=10 ELSE 260
220   IF ABS(ADV%-2048)<20 THEN GAIN%=100 ELSE 260

230   IF ABS(ADV%-2048)<2 THEN GAIN%=500
240 ' ***********************************
250 '
260   CALL XAV(CH%,GAIN%,ADV%)
270   VOLT!=(ADV%*20/4096-10)/GAIN%  ' convert ADV to Volts
280   PRINT "Voltage at channel";CH%;"(gain ";GAIN%;") =";VOLT!;"volt."
290   SOU=ABS(VOLT!)*1000+50:SOUND SOU,2:GOTO 200 ' freq = mV + 50
300   END
310 '
1000 '***** INITIALISATION  *****
1010 XRD=111:XAV=3      ' entry points of routines used
1020 ID%=0:CH%=4:ADV%=0:GAIN%=1  'arguments of the routines
1030 CALL XRD(ID%)  ' reset and identify the ADC card
1040 RETURN
```

This short program illustrates the simplicity of ADC using the PCTHERM software. In actual practice the user may work with XMV and get the voltage with maximum resolution directly. However, the knowledge of the conversion of the ADC output to the actual voltage is necessary in many situations where XMV cannot be used. These include multichannel scanning, continuous and DMA operations, etc. The actual program, even in this case, is hardly 30 lines. The use of subroutines to separate out functional blocks has been demonstrated above by separating the INITIALIZATION operation.

12.4 Platinum Resistance Thermometry under GPIB

Hardware Required:

- GPIB-PC with Turbo PASCAL Handler
- Keithley 619 Electrometer/Multimeter™ installed as DMM619
- PT-100 Temperature Sensor

This exercise is designed to teach the use of a platinum resistance thermometer operating in the pulsed mode. In this experiment the sensor is energized only for the duration of the measurement in order to minimize local heating. This experiment involves the use of a Keithley 619 Electrometer/Multimeter, which requires secondary addressing. The three stages of the exercise are as follows.

Stage 1: Use IBIC to do the following:

1. Find the 619 multimeter and set its secondary address to 61h
2. Select the resistance modes of the DMM
3. Make resistance measurements through the IBREAD command
4. Select the voltage mode of the DMM

Stage 2: Write a program to operate the DMM in the resistance mode and measure and display the temperature every second. The resistance of the sensor is 100 ohms at 0°C and increases by 0.3927 ohms per degree C. This will involve stripping the suffix from the data transferred by the DMM and the conversion of strings to numbers.

Stage 3: Operate the sensor in the pulsed mode and display the temperature at one second intervals at a fixed position in the middle of the screen. Hitting the <ESC> key or any function key should lead to an exit from the program.

The easiest way to achieve pulsed mode is to keep the multimeter in the voltage mode when not actually measuring resistance. This removes the excitation except when required, minimizing local heating.

The Turbo PASCAL program for stage 3 is given below:

program DMM_DEMO;

```
(*      KEITHLEY 619 ELECTROMETER/MULTIMETER OPERATION
          Requires Secondary Addressing
        Function
               -  Temperature Measurement with PT100 sensor
                  in the pulsed mode

        Sensitivity of PT 100  = 0.3927 ohms / deg C  *)

  {$I TPDECL.PAS}  { Include Turbo Pascal declarations}

  type tempstr=string[255];
  var code,i,id:integer;
      st1,st2:string[5];
      inwrd:iobuf;
      ch:char;
      v:real;            value of DMM output
      twrd:string[11];

   procedure LOADSTR(var outst:iobuf;instr:tempstr;len:integer);
    (* Function    :    Takes in a string and loads it in an array   *)
      var i:integer;
      begin
        for I:=1 to len do
          outst[i]:=instr[i];
      end;

   procedure SBWRT(dev:integer;message:tempstr;len:integer);
    (* Function    :  Takes in the command as a string argument and
                      sends it to the device through IBWRT function *)
     var wrt:iobuf;
     begin
       loadstr(wrt,message,len);
       ibwrt(dev,wrt,len);
     end;
```

```
procedure init;
(* Function      : To initialize the multimeter
                       GPIB Commands  Z0  -  No zero correction
                                      N0  -  No baseline storage
                                      U0  -  No baseline suppression
                                      P1  -  Filter ON
                                      R0  -  Autoranging
                                      S1  -  No averaging, 16.66s int.
                                      T1  -  One shot trigger on talk
                                      C0  -  Zero Check Disabled
                                      X   -  Execute     *)

  begin
    id:=ibfind('dmm619');              { identify the DMM }
    ibsad(id,$61);                     { send the secondary address }
    ibclr(id);                         { reset the DMM }
    ibloc(id);                         { send it to local mode }
    sbwrt(id,'C0X',3);
    sbwrt(id,'Z0N0U0P1R0S1T1X',15);    { send the initialization string }
  end;

  begin
    init;
    clrscr;
    gotoxy(14,1);
    write(' Platinum resistance thermometer in the pulsed mode');
    gotoxy(10,23);
    write('          Press any function key to stop');
    repeat
      ch :=chr(0);
      sbwrt(id,'F2X',3);       {select RESISTANCE mode through F2}
      ibrd(id,inwrd,18);        { read the resistance as an ASCII string }
      sbwrt(id,'F0X',3);       { select VOLTS mode through F0}
      gotoxy(30,12);
      for i:= 6 to 16 do
        twrd[i-5]:=inwrd[i]; { load only the numeric part of the input
                               string into another string              }
      twrd[0]:=chr(11);
      val(twrd,v,code);
      v:=(v-100)/0.3927;
      write(v:4:1);write(' C');
      delay(100);
      if keypressed then
        read(kbd,ch);
      if ch = #27 then exit;     {Exit on <ESC>}
     until ch = #27;
     ibloc(id);
   end.
```

12.5 Platinum Resistance Thermometry using DT2805

Hardware required:

- DT2805 card with PCTHERM software
- PT-100 Sensor
- 10 kohm precision resistor

The three stages of this exercise may be set up in a manner analogous to those in the Exercise 12.3. The connections are as per Fig. 12.2. The program for stage 3 is given below. The program measures the room temperature and PT-100 temperature every three seconds and displays them. For the PT-100 an average of twenty readings is taken. It should be noted that during the initialization the output voltage is set to 0, since the output of a DAC at power up will correspond to an analog data value of 0, which, in turn, will correspond to minus full range. Thus, for bipolar outputs it will not be zero.

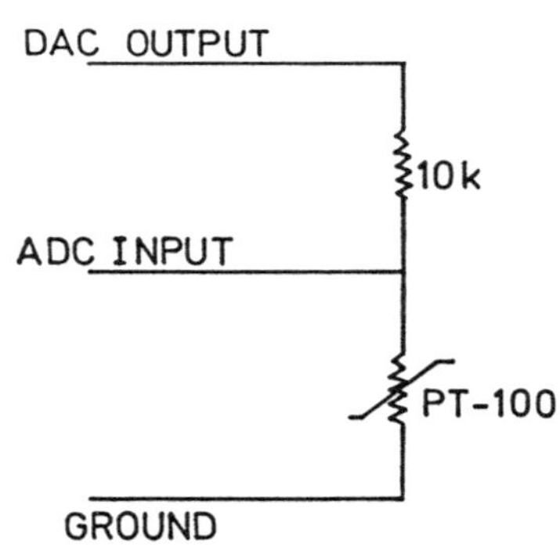

Figure 12.2—PT-100 Connections

```
10 '*============================================================*
20 '*                                                            *
30 '*   DEMONSTRATING MEASUREMENT OF TEMPERATURE USING PLATINUM  *
40 '*                RESISTANCE IN PULSED MODE                   *
50 '*                Programmer ..., Date ......               *
60 '*                                                            *
70 '*============================================================*
80 '  uses DT2805 card with PCTHERM
90 '  Requires Advanced Basica 2.0 and Min. 128kB RAM
```

```
100 '  Assumes PCTHERM.BAS to be in current directory
110 '
120 PRINT "loading PCTHERM... "
130 DEF SEG = 7552
140 BLOAD "pctherm",0
150 PRINT "initializing.....":GOSUB 350
160 GOSUB 420
170 TINIT=ET=0
180 CALL XMCJ(ROOMT!)
190 VOLT!=7!:CALL XGV(DAC%,VOLT!)   ' generate 7.0V at DAC
200 V.AT.R=0
210 FOR I%=1 TO 20
220 CALL XMV(CH%,VR!)               ' measure voltage at PT100
230 V.AT.R=V.AT.R+VR!
240 NEXT I%
250 V.AT.R=V.AT.R/20
260 R=10000*V.AT.R!/(7!-V.AT.R!)    ' Resistance used is 10kj
270 T=(R-100)/.3927
280 VOLT!=0:CALL XGV(DAC%,VOLT!)
290 LOCATE 13,17:PRINT USING "####.##";T
300 ON KEY (1) GOSUB 510
310 FOR I=1 TO 1000:NEXT
320 GOTO 180
330 SCREEN 0,0,0:KEY ON:END
340 '
350 '***** INITIALISATION  *****
360 DEFINT X
370 XRD=111:XMCJ=144:XMV=138:XGV=165   'entry points of routines used
380 ID%=0:ROOMT!=0:CH%=7:VR!=0:DAC%=0:VOLT!=0  'arguments of routines
390 CALL XRD(ID%)           ' reset and identify the DT2805 card
400 RETURN
410 '
420 '***** SCREEN FORMAT *****
430 KEY OFF:SCREEN 1:KEY (1) ON
440 LOCATE 19,7:PRINT "HIT <F1> TO TERMINATE"
450 LOCATE 5,7:PRINT "TEMPERATURE MEASUREMENT"
460 LOCATE 6,6:PRINT "USING PLATINUM RESISTANCE"
470 LOCATE 13,5:PRINT "Temperature":LOCATE 13,24:PRINT " (deg.C)"
480 RETURN
490 '
500 '***** SERVICING <f1> *****
510 WHILE INKEY$<>"":WEND:RETURN 330
```

12.6 Stepper Motor Driver

Hardware Required:

- DT2805 card with PCTHERM and/or DT2805 software drivers
- 4-pole stepper motor
- 4 solid-state relay TTL inputs and outputs suitable for the stepper motor
- 4 diodes for protection of relays (if not built in)
- Power supply suitable for stepper motor

This exercise will illustrate the use of digital outputs to drive a stepper motor. Stepper motors are becoming increasingly popular as actuators for rotary motion. Listings in both BASIC using PCTHERM and Turbo PASCAL using the DT2805.PAS driver are given.

Stage 1: Write a program to use the digital interface of the DT2805 as an output. Check that specific bytes can be output.

Stage 2: Connect the stepper motor to the digital IO lines according to the circuit in Fig. 8.11. and ensure that the connections are correct.

The connections can be checked as follows:

1. To ensure that all the relays are working and there are no major faults in the wiring, energize one pole at a time by sending the bit patterns 0000, 0001, 0010, 0100, and 1000. The motor should be free to rotate with all the poles off and relatively difficult to rotate if any of the poles are switched on. In case of problems the circuit needs to be checked.

2. Engergize the poles in the sequence 1100, 0110, 0011, and 1001 (Table 8.1). It may happen that in some of these positions the motor is free, while it is locked in others. If so, a logic table will have to be constructed and then some of the pole assignments may have to be changed.

It is essential to ensure proper connections before proceeding to Stage 3.

Stage 3: Write a program to move the motor in either direction as required. There should be provision for continuous rotation as well as by a prespecified number of steps in either direction as required.

The BASICA program is given below.

```
10 '===============================================================*
20 '*                                                              *
30 '*       EXERCISE TO DEMONSTRATE STEPPER MOTOR CONTROL          *
40 '*                      Name,    Date                           *
50 '*                                                              *
60 '*==============================================================*
80 '  uses DT2805 card with PCTHERM
90 '  Requires Advanced Basica 2.0 and min. 128 kB RAM
100 '  Assumes PCTHERM.BAS to be in current directory
105 '
110 KEY OFF:CLS:PRINT "Loading PCTHERM... "
120 DEF SEG = 7552
130 BLOAD "PCTHERM",0
140 GOSUB 1000
150 FOR L%= 1 TO 60:PRINT "-";:NEXT
155 PRINT "NO. OF STEPS TO RUN (0 or <CR> to run indefinitely)";
170 INPUT STEPS%
180 IF STEPS%<0 OR STEPS%<>INT(STEPS%) THEN BEEP:PRINT "Not valid. Try again ";:GOTO 170
190 PRINT "In forward or reverse direction (F/R)";
210 INPUT DN$
220 IF DN$="f" OR DN$="F" THEN 240
230 IF DN$="r" OR DN$="R" THEN 320 ELSE SOUND 100,2:PRINT "- Give F or R only -";:GOTO 210
240 GOSUB 2000:J%=1
250 WHILE K%<>STEPS%
260 CALL XODV(PORT%,MASK%,A%(J%))
270 ON KEY (1) GOSUB 4000
280 K%=K%+1:J%=J%+1
290 IF J%=5 THEN J%=1
300 WEND
310 GOTO 390
320 GOSUB 2000:J%=4
330 WHILE K%<>STEPS%
340 CALL XODV(PORT%,MASK%,A%(J%))
350 ON KEY (1) GOSUB 4000
360 K%=K%+1:J%=J%-1
370 IF J%=0 THEN J%=4
380 WEND
390 PRINT "Specified (";STEPS%;") steps completed.";
400 SOUND 400,3:PRINT " (Motor holding) "
410 PRINT "HIT <ESC> TO EXIT, any other key to begin again..."
420 E$=INKEY$:IF E$="" THEN 420
430 IF ASC(E$)=27 THEN 440 ELSE 150
440 FOR L%= 1 TO 60:PRINT "-";:NEXT:PRINT
450 CALL XRD(ID%):PRINT "Motor Released."
460 KEY ON: END
```

```
1000 '----- INITIALISATION ------
1010 DEFINT X
1020 XRD=111:XEFO=60:XODV=66
1030 ID%=0:PORT%=1: MASK%=&HF
1040 DIM A%(5)
1050 A%(1)=&HC: A%(2)=&H6: A%(3)=&H3: A%(4)=&H9: A%(0)=0
1060 CALL XRD(ID%)
1070 CALL XEFO(PORT%)
1080 CALL XODV(PORT%,MASK%,A%(0))
1090 RETURN
2000 '------ PRE-RUN -------
2010 IF STEPS%=0 THEN STEPS%=-1
2020 K%=0: KEY (1) ON:PRINT
2030 PRINT "RUNNING...        [ HIT <F1> TO INTERRUPT ]"
2040 ON ERROR GOTO 3000
2050 RETURN
3000 '---- OVERFLOW TRAP ------
3010 K%=0: RESUME
4000 '---- <F1> SERVICE -----
4005 WHILE INKEY$<>"":WEND  'Flush keyboard buffer
4020 PRINT:PRINT "INTERRUPTED *** ";
4030 RETURN 400
```

The Turbo PASCAL program using the DT2805.PAS handler is given in Section 12.7. It may be noted that a delay of 5 ms between successive steps has been introduced to give the stepper motor sufficient time to respond. This is necessary due to the higher speed of the compiler, as compared to the interpreter.

```
Program Stepper;
(*      Program to drive stepper motors in either direction
        The motor can run for prespecified number of steps or indefinitely
        Program developed by
                                Programmer
                                Date mm dd yy                    *)
{$I DT2805.PAS}

const
      Outport = 1;          {Motor is connected to port 1}
      Mask = $0F;           { and bits 0 to 3  }
 var
     Index,Dum,Steps:integer;
     C,Direction:char;
     Seqloc:array [0..3] of byte;

  procedure Drawline;
  begin
    writeln('-------------------------');   end;
```

```
procedure Initialize;
  var dummy:integer;

  begin
    clrscr;
    xrd(dummy);                        {Initialize the DT2805 card}
    Seqloc[0]:=$C;                     {Set up the byte sequence  }
    Seqloc[1]:=$6;
    Seqloc[2]:=$3;
    Seqloc[3]:=$9;
    xefo(outport);                     {Make OUTPORT the output port}
    xodv(outport,mask,0);
    Direction:='F';                    {Default direction is forward}
    Steps:=-1;                         {Default # of steps - indefinite}
    Index:=1;
  end;

procedure Setupparam;
  var x,y:integer;
      newdirection:char;
  begin
    drawline;
    write('No. of steps to run (0 or <CR> to run indefinitely) ');
    x:=wherex;y:=wherey;
    repeat                             {Keep asking till a numeric input}
      gotoxy(x,y);clreol;              {is provided.                    }
      {$I-} readln(steps); {$I+}
    until ioresult  = 0;
    if steps = 0 then steps:=-1;
    write('In forward or reverse direction(F/R)? ');
    repeat
      read(kbd,newdirection);
      newdirection:=upcase(newdirection);
      if not (newdirection in ['F','R']) then  {If input not right, hoot}
        write(^G);
    until newdirection in ['F','R'];
    if newdirection<>direction then
       if newdirection = 'F' then index:=(index+1) mod 4
                             else index:=(index+3) mod 4;
    direction:=newdirection;
    writeln(direction);
    writeln;
    writeln('Running ...            [Hit F1 to stop]');
    writeln;
   end;
```

```
  procedure Runmotor;
    var Countloc:integer;
        Outbits:byte;
        ch:char;
    begin
      Countloc:=0;

      while Countloc <> steps do
        begin
          case Direction of
            'F':begin
                  xodv(Outport,Mask,Seqloc[Index]);
                  Index:=(Index+1) mod 4;
                end;
            'R':begin
                  Index:=(Index+3) mod 4;    {Access in reverse sequence}
                  xodv(Outport,Mask,Seqloc[Index]);
                end;
          end;
          if Countloc = Maxint then
            Countloc:=0
          else
            Countloc:=Countloc+1;

          if keypressed then              {Check for F1 key }
            begin
              read(kbd,ch);
              if (ch = #27) and (keypressed) then
                begin
                  read(kbd,ch);
                  if ch = #59 then
                    begin
                      writeln('             **Interrupted!');
                      exit;
                    end;
                end;
            end
          else
            delay(5);
        end;
        writeln('Specified (',steps,') steps completed.');
        writeln('(Motor holding)');
    end;

begin
  Initialize;
  repeat
```

```
    Setupparam;
    Runmotor;
    writeln('Hit ESC to exit, any other key to begin again...');
        read(kbd,c);
        writeln;
        if (c = #27) and (not(keypressed)) then
          begin
            xrd(dum);
            drawline;
            writeln('Motor released');
            halt;
          end;
  until Steps <> Steps;
end.
```

12.7 Writing a Device Handler

This section discusses the writing of a device handler in Turbo PASCAL. The handler is very similar to the PCTHERM handler, which is supplied by Data Translation for the DT2805 interface cards. The Data Translation cards have on-board processors and communicate to the host PC through a PIA occupying two successive addresses in the input-output map. In this case the base address is 2ECh and can be changed by changing the constant declaration. The masking is of the simple AND type, discussed in Chapter 8.

The handler does not use DMA or allow for external triggering, etc. The timer internal to the DT card can not be programmed either. It simply illustrates the way in which the interface can be driven through a high-level language and should serve for many simple experiments. The commands supported are Reset (XRD), Stop (XSTOP), Enable for output (XEFO), Enable for input (XEFI), Output digital value (XODV), Set-up ADC (XSA), ADC series (XAS), and Measure temperature for K-type thermocouple (XMV). Examination will show that extensive use of the Turbo PASCAL port access command PORT is made. Reference to the DT280x manual will reveal that once the port access routines XERROR, XCOMMAND, XREADY, X_DATA_IN_EMPTY, X_DATA_OUT_READY, and READ_BYTE are written, the rest of the handler is easily developed using the Commands section (Section 3) of the Data Translation DT Cards manual. All but the last routines are simply reading the status byte and checking prespecified bits to get the information.

```
(*********************************************************)
(*          Driver for DT2805 ADC/DAC card              *)
(*                        Name,    Date                 *)
(*                                                      *)
(*********************************************************)
```

```
  type datarray=array[1..16] of integer;
  const base address=$2EC;              {base address of DT2805}
  var data_in,data_out,command,status:integer;

  function XERROR:boolean;                 {error bit}

(* This checks the status byte of the DT2805 card.  If an error occured in
the last operation with the card it set to TRUE otherwise it returns a value
FALSE        *)

    var t:byte;
    begin
      t:=port[status];
      xerror:=((t and $80)=$80);               {Check status 7 bit}
    end;

  function XCOMMAND:boolean;                    {command bit}

(*  This function checks the status byte to see if the card has received the
last byte as command or data *)

    var t:byte;
    begin
      t:=port[status];
      xcommand:=((t and $08)=$08);         {Check status 3 bit}
    end;

  function XREADY:boolean;                      {ready bit}

(* This function checks the status byte to see if the card has completed the
last operation.  If the last operation is over then it returns a value TRUE
otherwise FALSE             *)

    var t:byte;
    begin
      t:=port[status];
      xready:=((t and $04)=$04);             {Check status 2 bit}
    end;

  function X_DATA_IN_EMPTY:boolean;             {data in full bit}

(* This function checks the status byte to find if the input register is
empty.  If it is empty, the function returns a value TRUE otherwise it
returns a value FALSE             *)

    var t:byte;
    begin
```

```
      t:=port[status];
      x_data_in_empty:=not((t and $02)=$02);  {Check status 1 bit}
    end;

  function X_DATA_OUT_READY:boolean;          {data out redy bit}

(* This function checks the status byte to find if the output register is
ready to output data.  If it is empty, the function returns a value TRUE
otherwise it returns a value FALSE*)

    var t:byte;
    begin
      t:=port[status];
      x_data_out_ready:=((t and $01)=$01);    {Check status 0 bit}
    end;

   function READ_BYTE:byte;

   (* This function returns the byte in the data out register    *)

     begin
       repeat until x_data_out_ready;
       READ_BYTE:=port[dataout];
     end;

  procedure XSTOP;

 (* This procedure stops the operation in progress     *)

    var t:byte;
     begin
       port[command]:=$0F;
       t:=port[data_out];
     end;

  procedure XRD(var t:integer);

 (* This routine initializes the DT2805 card for further operations.  First
 of all the IO port addresses are set up from the base address.  Then the
 card is given a RESET  *)
      var dummy:byte;
      begin

        {Set up the IO addresses}
        data_in:=base_address;
        data_out:=base_address;
        command:=base_address+1;
```

```
      status:=base_address+1;

      {Now initialize}
      xstop;
      dummy:=port[data_out];
      repeat until xready;
      if xerror then
      begin
        port[command]:=$01;
        repeat until xready;
      end;
      port[command]:=$00;
      repeat until x_data_out_ready;
      t:=port[data_out];
      repeat until x_data_in_empty;
    end;

procedure XODV(portno:byte;mask,val:integer);

(* This routine outputs a digital value to the a selected port with a given
mask.  The selected port comes in portno as a byte, the mask to be used
for the output and the value to be outputed comes in the variables mask
and val.      *)

  var t:byte;
      t1:integer;
  begin
    repeat until xready;
    port[command]:=$07;          {write Digital output immediate}
    repeat until x_data_in_empty;
    port[datain]:=portno;
    t1:=mask and val;             {perform masking}
    repeat until x_data_in_empty;
    t:=t1;
    port[data_in]:=t;             {write masked value to port}
    if portno=$02 then            {write the high byte}
      begin
        t:=t1 div 256;
        repeat until x_data_in_empty;
        port[data_in]:=t;
      end;
    repeat until x_data_in_empty;
  end;

  procedure XEFO(portno:byte);
```

```
(* This routine enables the port chosen in the variable portno for output *)

    var t:byte;
    begin
      repeat until xready;
      port[command]:=$05;                      {EFO command}
      repeat until x_data_in_empty;
      port[data_in]:=portno;
      repeat until x_data_in_empty;
      xodv(portno,0,0);
    end;
  procedure XEFI(portno:byte);

(* This routine enables the port chosen in the variable portno for input *)

    var t:byte;
    begin
      repeat until xready;
      port[command]:=$04;                      {EFI command}
      repeat until x_data_in_empty;
      port[data_in]:=portno;
      repeat until x_data_in_empty;
    end;

  procedure XSA(source,channel_st,channel_fin,gain:integer);

  (* This routine sets up the various parameters of the ADC.  The variable
source is not used, but is intended to be used for setting up the kind of
timming source required.  The varibles channel_st,channel_fin gets the start
and the finishing channel numbers to be scanned and the variable gain gets
one of the possible gains to be used for the channels used.          *)

    var t:byte;
    begin
      repeat until xready;
      port[command]:=$0D;
      repeat until x_data_in_empty;
      case gain of                             {set up the gain}
        500: port[data_in]:=$03;
        100: port[data_in]:=$02;
         10: port[data_in]:=$01;
          1: port[data_in]:=$00
      end;
      repeat until x_data_in_empty;
      t:=channelst;
      port[data_in]:=t;
      repeat until x_data_in_empty;
```

```
      t:=channel_fin;
      port[data_in]:=t;
      repeat until x_data_in_empty;
      port[data_in]:=$03;            {dummy no. of conversions}
      repeat until x_data_in_empty;
      port[data_in]:=$00;                     {high byte}
      repeat until x_data_in_empty;
    end;

procedure XAS(no_of_values:integer;var values:datarray);

(* This routine reads the ADC continously. The variable no_of_values stores
the number of convertions required, and gives the read values in the array
variable values. This variable is 16 locations (of integer type) long. If
no_of_values is more than 16 change this length in the global decleration of
dataarray accordingly.            *)

  var l,h:byte;
      i:integer;
  begin
    repeat until xready;
    port[command]:=$2E;                      {Read A/D continous}
    for i:=1 to no_of_values do
      begin
        l:=read_byte;
        h:=read_byte;
        values[i]:=l+256*h;
      end;
    xstop;                                    {stop conversions}
  end;

procedure sort(var nums:datarray;num:integer);

(* This routine does the quick sort of the values in the variable nums and
the sorted values are put back in the same variable. The sorted values are
in the ascending order. *)

    var temp1:integer;
        k,m,i,j,pass,gap:integer;

    procedure swap;
      begin
        temp1:=nums[j];
        nums[j]:=nums[j+gap];
        nums[j+gap]:=temp1;
      end;
```

```
   begin
     for k:=1 to num do
     begin
        gap:=num div 2;
        while gap>0 do
          begin
            for i:=(gap) to num do
              begin
                j:=i-gap;
                while j>0 do
                  begin
                    if nums[j]>nums[j+gap] then
                      begin
                        swap;
                        j:=j-gap;
                      end
                    else j:=0;
                  end;
                end;
                gap:=gap div 2;
          end;
      end;
  end;

procedure XMT(tctype,channel:integer;var deg_C:real);

(* This function reads the channel indicated by the variable channel and
gives the equivalent temperature in the variable degC.  The variable tctype
is not used but is intended to be used to specify the type of thermocouple
used to measure the temperatue. Currently only type K Chromel-Alumel is
supported. The routine has the capabilities to set up the required gain
automatically  if the requirment lies in the range of >=100 and <=500. For
more ranges, the setgain routine will have to be modified.          *)

label 100;
var cdata:array[1..2,1..2] of real;
    tcv,rt,sumav:real;
    index:integer;
    temp:datarray;

  procedure setgain(chn:integer);

(* This routine is used by XMT to auto set the gain of the channel number
chn. The data required for each gain is in the global variable cdata.  The
logic used is to start with the minimum gain and take a reading.  Keep
switching to the next higher gain till the reading taken is <= to 80% of
the full scale reading of the current gain. *)
```

```
    begin
      xsa(0,chn,chn,100);
      index:=2;
      xas(1,temp);
      tcv:=cdata[2,index]+cdata[1,index]*temp[1];
      if tcv>abs(0.8*(4095*cdata[1,index]+cdata[2,index])) then
        begin
          xsa(0,chn,chn,500);
          index:=1
        end;
    end;

  procedure xmt1;

(* Routine used by XMT to get the average value of 16 reading after
filtering. The median 8 of the sorted values are used to get the average
value.          *)

  var temp:datarray;
      sum,i:integer;
    begin
      xas(16,temp);
      sort(temp,16);
      sum:=0;
      for i:=5 to 12 do
          sum:=sum+temp[i];
      sumav:=sum/8.0;
    end;

begin
  cdata[1,1]:=9.7656E-03;  cdata[2,1]:=-20.00;        {gain 500}
  cdata[1,2]:=48.828E-03;  cdata[2,2]:=-100.0;        {gain 100}
  setgain(0);
  xmt1;
  rt:=(cdata[2,index]+cdata[1,index]*sumav)*2.0;  {0.5mV per °C}
  if channel=0 then
    begin
      deg_C:=rt;
      goto 100;
    end;
  setgain(channel);
  xmt1;
  tcv:=cdata[2,index]+cdata[1,index]*sumav;  {mV output}
  deg_C:=tcv/0.041269 + rt;             {1000°C = 41.269 mV}
  100:
end;
```

It is hoped that the exercises in this chapter have given the reader some insight into the techniques of PC interfacing and that the sample programs have provided some guidance in the ways of successfully writing working software.

Suggested Reading

As the hardware available in individual laboratories varies, the user should consult the manuals of the equipment available. For reading material on interfacing, kindly refer to Chapters 6 through 10.

Glossary

actuator: A device used to operate a control element.

analog data value: The digital value corresponding to an analog signal. (See Chapter 8.)

analog-to-digital converter (ADC): An electronic device used to convert an analog signal to the digital form, which may then be sent to the computer.

asynchronous: A data transfer process where the source and destination are not synchronized. Asynchronous transfers are used by most communication schemes, like RS232C, Centronics etc.

ASYST: A data analysis and acquisition package marketed by Macmillan Software. ASYST is one of the most powerful integrated systems available, and is characterized by a FORTH-like syntax.

BASIC (BASICA, GWBASIC, QBASIC): Beginners All Purpose Symbolic Instruction Code. A simple and popular computer language used extensively. BASICA and GWBASIC are the advanced versions of the Interpreter often shipped as a part of PCDOS and MSDOS, respectively. QBASIC is the newest (and structured) manifestation.

baud: A line modulation rate which gives the highest signalling rate for the medium. It is the inverse of the shortest signal element (in seconds). Should not be confused with the bit rate as only on baseband (DC) non-encoded signal will the baud rate equal the bit rate. Common baud rates are 300, 600, 1200, 2400, 4800, 9600 and 19200. It should be noted that a standard telephone line has a nominal 1200 baud capacity. Using encoding techniques modems are able to operate at far higher speeds (14.4 kbps is common) by fooling the line to believe that there are only 1200 changes per second.

bidirectional port: A communication port which can function as both input and output.

BIOS: Basic Input Output System. The BIOS is the nucleus of the computer operating system.

bit: Binary digit. A binary number, i.e., 1 or 0. Computers use binary numbers to express all numbers.

buffer: A temporary storage used in data communication. During transfers from one device to another the buffers are used to temporarily store data until the receiving device is ready to process it.

bus: Used to denote either i. a communication network topology, or ii. part of the architecture of the computer.

byte: 8-bits. The unit of computer data handling.

Centronics interface: A very popular parallel interface used for connecting printers to computers.

Cheapernet: Same as 10 Base 2.

client-server: A distributed system, where the server and the client each perform part of the task. Most PC based client-server systems comprise a networking situation where one (or a few) computers act as hosts for other systems. The expensive resources (i.e., hard disks, printers, etc.) are concentrated in the servers and are used by all other systems (the clients). Usually the servers are dedicated for the purpose and not used for computing.

color graphics card (color graphics adapter, CGA): A standard graphics card for the PC. Has two text modes of operation—40 char. × 24 lines, and 80 char. × 24 lines. There are two graphics modes—320 × 200 pixels with 16 colors, or 640 × 200 in two colors.

COM: Logical name for serial (COMmunication) port.

compiler: Software that translates a high level language program (source code) into the machine-readable form.

conditioning: Processing (e.g., filtering, amplifying, etc.) of a signal to make it acceptable for further handling.

controller: An add-on hardware (card) that enables software control of (the attached) device it is controlling. For example, a floppy disk controller, or GPIB controller.

co-processor: A specialized device which has been specially designed for fast processing of a specific type of data. The co-processor takes over the processing of data appropriate to it from the main CPU and does so many times faster. For example, in the PC the 8087 Numeric Co-processor, if present, can process numerical data up to 30 times faster than the 8088 can. In order exploit the speed of a co-processor the software must be able to take advantage of it.

CSMA/CD: Carrier Sense Multiple Access/Collision Detect. A protocol used in local area networking (Ethernet) to allow many units to share the same data pathway.

current loop: A popular way of operation of process control instruments. The current flowing in the circuit is proportional to the signal. Current loop operation offers good noise immunity and long cables do not effect the signal value.

data structure: The organization of the data in any computer application. (See Chapter 2.)

device driver: Software used to integrate the software required to communicate with a device into the DOS. A device driver is usually installed through a DEVICE=... command in the CONFIG.SYS file. Occasionally it may take the form of a resident program.

device handler: It is different from a device driver. Software which is used as a part of user programs to communicate with a device. A device handler is not installed as a part of DOS. A software handler is the software for linking the user software to the device driver.

digital-to-analog converter (DAC): A hardware device used to convert digital (discrete) signals into the continuous (analog) form.

direct memory access (DMA): The process by which data can be directly transferred to or from the memory to a peripheral unit bypassing the main CPU. DMA transfer of blocks of data is much faster than normal transfer through the CPU. The PC has four DMA channels out of which two (one on the XT) are available to the user. AT systems provide three additional DMA channels.

DOS: Disk Operating System. The operating environment of the computer. In the context of the PC the term DOS is also used to refer to MSDOS or PCDOS.

dot matrix printer: A printer which forms images by printing closely spaced dots. Dot matrix printers are fast, cheap and can do graphics. However, they cannot match the quality of letter quality printers for text and plotters for figures. Laser printers are also dot matrix printers with a very high resolution, matching the letter quality printers and plotters, and provide all the flexibility of dot matrix devices.

dynamic signal: A signal which varies with time.

editor: A program used for modifying text files. It is more specifically used for software designed for editing ASCII files of data or programs.

EISA: Extended Industry Standard Architecture. An extended version of the PC-AT bus, developed jointly by nine major PC manufacturers as an alternative to the Micro Channel Architecture of the PS/2 range.

enhanced graphics adapter (EGA): The Enhanced Graphics Adapter or Card typically offers 640 × 350 resolution in 16 colors. Most EGA cards can emulate CGA as well as Hercules cards.

EPROM: Erasable Programmable Read Only Memory. EPROMs are used to store the firmware of computers. Once programmed they retain information until erased by exposure to ultraviolet light. The BIOS, etc. are stored in EPROMs in most computers.

Ethernet: A popular LAN using the CSMA/CD protocol.

firmware: Software which forms part of the computer and is seldom if ever changed, e.g., the BIOS, character generator, etc. Firmware is installed into the system in the form of ROM.

FORTH: FOuRTH generation language. A language designed for process control applications, working on the reverse polish notation, in which the operator follows the operand. It is ideally suited for stack based machines, and is very popular in control applications in astronomy.

FORTRAN: FORmula TRANslator. The first general purpose scientific computing language, and still the most popular.

GPIB: General Purpose Interface Bus. The common name used for the IEEE 488 protocol. (See Chapter 7.)

grounding: Connecting the various voltage points to a common point. Bad grounding practice is the most common source of noise in measurements.

GUI: Graphical User Interface. A user friendly, intuitive interface that uses graphical techniques and icons for communicating with the user.

handler: Software for using an interface device. See Device Driver, Device Handler.

handshaking: A method used in data communications to interlock the transmitter- and receiver, in order to ensure efficient data transfer, without loss of data, or unnecessary delays. (See Chapter 6.)

hardware: The various electronic (and mechanical) components used in a computer.

Hercules Graphics Card (HGC): A display card developed by Hercules, which emulates the IBM Monochrome Adapter for text, and can also be used with a 720 × 348 resolution for graphics. This is the only non-IBM graphics card which is almost universally supported by software. Clones of the Hercules card are often called Monochrome Graphics Cards (Adapters).

HPGL: Hewlett Packard Graphics Language. A plotter control language developed by Hewlett Packard, which is the most popular plotter control language.

HPIB: Hewlett Packard Interface Bus. Trade name used by HP for GPIB. HPIB protocols are very highly standardized interpretations of the IEEE 488 standard.

IDE: Integrated Drive Electronics. The popular hard disk interface in which the controller is built into the drive itself and requires only a simple interface to the PC-AT bus.

IEC625: International Electronics Commission's Specifications for the IEEE 488 standard. It differs from IEEE 488 primarily in the connectors used.

IEEE 488: The Institution of Electrical & Electronics Engineers Specifications for GPIB, HPIB or IEC625.

industrial PC: A ruggedised version of the PC designed to cope with the harsh shop floor environment. (See Chapter 1.)

in-line code: A technique used in Turbo PASCAL to allow the use of machine language instructions inside a program.

instrumentation amplifier: High gain, low noise amplifiers used in instrumentation. (See Chapter 5.)

interface: A system which allows the interconnection of two dissimilar devices.

interpreter: A language translator, which stores the program as entered and then converts each line of code to machine language as it executes it. (See Chapter 6.)

interrupt: A technique used to signal the occurrence of a random event to the computer. Interrupts allow the fast handling of such events, but require additional hardware and software.

IO addressing: A method in which input-output takes place in a different address space than that used by the memory.

ISA: Industry Standard Architecture. Another name for the PC-AT bus.

isolation: Electrical separation of two circuits. Isolation protects the two systems from mutual electrical interference. It is often used to connect a high voltage or power stage to the controller which may otherwise be adversely affected by transients and leakage signals.

I to P converter: An electromechanical device used in controls to convert a current loop signal (usually 4–20 mA) to a pressure signal to operate the actual control element.

LAN (Local Area Network): The interconnection of computers over a short distance with a view of sharing data, resources, and permit digital communication. (See Chapter 6.)

listener: A device configured to receive signals or data in GPIB interfacing. (See Chapter 7.)

Lotus 1-2-3: A very popular spreadsheet developed by Lotus Development Corporation.

LPT: Logical name for a parallel printer port.

MAP: Manufacturing Automation Protocol. A LAN first developed by General Motors for shop floor automation. It carries the IEEE Standard No. 802.4. (See Chapter 6.)

masking: An IO technique in which only selected bits in the input can affect the output. (See Chapter 8.)

monochrome graphics card (MGC): Same as Hercules card.

motherboard: The printed circuit board to which all other feature cards, power supply, keyboard, etc., are connected.

mouse: A pointing device used mostly with GUIs.

MSDOS: Microsoft Disk Operating System. The standard operating system for the PC family. The version used by IBM is called PCDOS. There are only minor differences between PCDOS and MSDOS.

multiplexing: Sharing of a signal path by many signals.

noise: Unwanted signal.

numeric coprocessor: See Co-processor.

OOP: Object Oriented Programming. Modular programming in which program and data structures are broken up into objects that can inherit characteristics from their parents and pass them on to the children. A logical extension of modular programming.

operating system: The operating environment of a computer. (See Chapter 3.)

parallel interface: An interface in which all the bits of data are presented at the same time. (e.g., Centronics, GPIB.) (See Chapter 6.)

parallel interface adapter (PIA): Devices used in parallel connections.

parity: The extra bit used to confirm data integrity. In even parity the parity bit is set such that the data byte and the parity bit together have an even number of ones. The PC and most clones use parity.

PASCAL: A structured programming language developed by Nicklaus & Wirth. It is often the first programming language taught.

PC, XT, AT, PS/2: Various models in the PC family. (See Chapter 1.)

PCDOS: Personal Computer Disk Operating System. The IBM version of MSDOS.

PCI: Peripheral Component Interconnect. A high performance bus extension defined by Intel. Becoming increasingly popular in Pentium-based systems.

peer-to-peer: A networking situation in which all systems are equal, i.e., all (or many) computers can act as servers as well as clients for other systems. This is in contrast to the client-server system in which one (or at most a few) system acts as a dedicated server to the whole network.

Pentium: The Intel CPU that is a successor to the 80486. It uses a 64-bit data bus. Intel dropped the numerical designations because numbers could not be copyrighted.

polling: An IO technique in which the software checks the IO port periodically for data. Polled IO does not require any additional hardware but is slower and more inefficient than Interrupt driven IO.

port: A set of addresses to which an interface can be attached.

procedure: A logical unit of a program. It is the same as a subroutine in FORTRAN.

PROM: Programmable Read Only Memory.

RAM: Random Access Memory.

resolution: The fineness with which an input can be represented or visualized.

ROM: Read Only Memory. (See also RAM, PROM, EPROM.)

RS-232C: The commonly used serial interface. It is used by a wide variety of devices like printers, plotters, modems, mice, etc.

sensor: A device used to detect a physical quantity and convert it to a measurable form. (See Chapter 5.)

serial interface: An interface in which data bits follow each other sequentially. It is used by most LANs, RS-232C, etc.

shielding: Protecting an instrument or cable from picking up (or radiating) extraneous signals or noise.

signal bandwidth: The frequencies between which the signal is usable.

signal conditioner: See Conditioning.

smart sensors: A sensor with an on board microprocessor to take care of signal conditioning, etc. Smart sensors offer multiple ranges, calibrations, etc.

software: Computer programs, etc., which reside in RAM during computer operation.

spreadsheet: A program used for working with tabular data, e.g., Lotus 1-2-3, Supercalc, Quattro, etc. In scientific work these are used for the analysis of tabular data.

static signal: A signal which does not vary with time.

stepper motor: A type of motor which moves in discrete steps. These usually have 60, 100, 200 or 400 steps per revolution. Stepper motors are specially suited for digital control, since the step size is fixed. Using special microstepping techniques resolutions upto 50,000 steps/revolution are possible. (See Chapters 8 and 12.)

subroutine: A self-contained procedure in FORTRAN. A subroutine cannot run by itself, it must be part of a larger program.

synchronous: A data transfer process where the transmitter and receiver are synchronized.

10 Base 2: The Thin Ethernet, also known as Cheapernet. The most commonly used cabling system based on 50Ω RG58A (or B)/U coaxial cable.

10 Base 5: The Thick Ethernet. Used as a backbone for large networks. The cable though still of 50Ω impedance is much thicker than in 10 Base 2, and can be tapped using transceivers only at specified points.

10 Base T: The Unshielded Twisted Pair Ethernet connection. This is in a star configuration with twisted pair cable using a modular jack similar (but larger) to the US telephone connector.

token passing ring: A common LAN protocol where the data transfer moves in a fashion akin to the game of pass the parcel. MAP is a token passing ring LAN. (See Chapter 6.)

top-down design: Software design process where the program is broken down into increasingly smaller segments prior to actual coding. (See Chapter 2.)

transient signal: A signal which is present for a very brief period of time.

transducer: A sensor or actuator.

talker: The device transmitting data in a GPIB network. (See Chapter 7.)

UNIX: An operating system developed by AT&T. It is becoming increasingly popular on mini- and larger systems. UNIX is also available for the AT and higher PCs.

VESA: Video Equipment Standards Association.

VGA: Video Graphics Array. A most popular graphics card that supports graphics on a 640 × 480 resolution. Super VGA modes include 800 × 600, 1024 × 768, etc. The VGA has a square pixel, i.e., a 10 × 10 box appears as square on the screen.

VL Bus: VESA Local Bus, an extension to the ISA (and EISA) bus, which provides a direct link to the CPU. Physically present as an additional connector on the board.

Winchester: A hard disk drive with non-removable media.

Windows: The popular graphical user interface developed by Microsoft. The term "window" is also used sometimes to define a region of the screen used for input or output.

word: The amount of data handled by the processor in a single step.

word processor: Software used for editing of documents, etc. (See also Editor.)

WYSIWYG: What You See Is What You Get. A characteristic of many text processing programs.

Index

About the Editors

1. **Dr. Sanjay Gupta** (Ph.D., University of Reading, U.K.)
   Presently Principal Scientific Officer, Advanced Centre for Materials Science, I.I.T. Kanpur. Research interests include process control, microcomputer applications in instrumentation, neutron scattering. Has done work at various neutron facilities, including Institut Laue Langevin, Grenoble, France; Rutherford Appleton Laboratory, Chilton, U.K.; the Atomic Weapons Research Establishment, Aldermaston, U.K.; and the Bhabha Atomic Research Centre, Bombay. Was Senior Research Fellow at the University of Reading from 1989–91. He has attended various conferences in U.K., U.S.A., Austria, Belgium, France, China, Singapore, and India.

2. **Dr. J. P. Gupta** (Ph.D., University of Pennsylvania, USA)
   Presently Professor, Chemical Engineering Department, I.I.T. Kanpur. Research interests include process control, microcomputer applications, energy conservation, chemical process equipment design, biotechnology, safety, and hazards. Worked earlier at Universal Oil Products Chicago; and University of Pennsylvania, Philadelphia. Was visiting professor at University of Mexico, Mexico City; University of Michigan, Ann Arbor; University of the South, Bhia Blanca, Argentina as a UNIDO expert. Was Science Councellor at the Indian Embassy, Washington DC from 1988–91. Attended various conferences in U.S.A., U.S.S.R., Japan, Yugoslavia, Singapore, Malaysia, Canada, and India. Member of several learned societies and on the Editorial Advisory Boards of several international journals. One book published by Hemisphere Pub. Co., N.Y. (Div. of Taylor & Francis, London).

A Brief Note about the Indian Institute of Technology, Kanpur

The Indian Institute of Technology (I.I.T.) at Kanpur was established in 1960 by an act of the Indian Parliament and was helped by a consortium of nine U.S. Universities (M.I.T., Stanford, Caltech, Berkeley, Purdue, Ohio State, Michigan, Carnegie, Case) for the first ten years in terms of visiting professors, planning and equipping the laboratories and library, and training of the faculty in the U.S. I.I.T. Kanpur is one of five I.I.T.s in India that are known as Institutes of National Importance. Its undergraduate engineering program is comparable to the best in the world. The five I.I.T.s select about 1,500 candidates out of a total of about 90,000 that take a national test for admission to these institutions. Graduates of I.I.T. Kanpur are to be found in 37 countries at last count and have made a name for themselves in numerous fields. The recruitment of faculty is very selective as well, and selections are done on global basis.

About the Authors

Sanjay K. Bose (Ph.D., State University of New York, Stony Brook, USA) is Professor of Electrical Engineering at the Indian Institute of Technology, Kanpur. His areas of interest include Computer Networking, Communication Theory, and Microprocessor applications. He is the author of *Digital System—From Gates to Microprocessors*, published by Wiley Eastern.

Sanjay Bhatnagar (M. Sc., M. Tech., I.I.T. Kanpur, India) is working on Image Processing at the GMRT Radio Telescope with the Centre for Development of Advanced Computers at Pune. He earlier worked in the Department of Metallurgical Engineering at I.I.T. Kanpur.

Sandeep Garg (B. Tech., I.I.T. Kanpur, India) has worked in broadcasting and telecasting media, as well as in software development. He is currently with the Department of Chemical Engineering at I.I.T. Kanpur, developing software for Hazard Management and Analysis.

Jai P. Gupta (Ph.D., University of Pennsylvania, Philadelphia, USA) is Professor of Chemical Engineering at the Indian Institute of Technology, Kanpur. His areas of interest include Computer Control, Transport Phenomena, Design, Chemical Safety and Hazards. He is the author of *Fundamentals of Heat Exchanger and Pressure Vessel Technology*, published by Hemisphere/Springer Verlag.

Sanjay Gupta (Ph.D., University of Reading, Reading, UK) is Principal Scientific Officer in the Advanced Centre for Materials Science at the Indian Institute of Technology, Kanpur. His interests include Instrumentation, Process Control, and Neutron Scattering.

Prem Malhotra (Ph.D., University of Waterloo, Canada) was Senior Research Engineer in the Advanced Centre for Electronics Systems at the Indian Institute of Technology, Kanpur. He is now working in Canada.

S. S. Prabhu (Ph.D., University of Toronto, Toronto, Canada) is Professor of Electrical Engineering at the Indian Institute of Technology Kanpur. His main research interest is the Application of Control Theory in Practical Areas.

Rajeev Sangal (Ph.D., University of Pennsylvania, Philadelphia, USA) is Professor of Computer Science and Engineering at the Indian Institute of Technology, Kanpur. He is the author of *Programming Paradigms in LISP*, being published by McGraw Hill. His areas of interest include Artificial Intelligence and Natural Language Processing.

V. C. Srivastava (B. Tech., I.I.T. Kanpur, India) is working in the Television Centre of I.I.T. Kanpur. He was earlier Research Fellow in the Advanced Centre for Material Science at the Indian Institute of Technology, Kanpur.